Lecture Notes in Artificial Intelligence   9767

Subseries of Lecture Notes in Computer Science

More information about this series at http://www.springer.com/series/1244

Brian Davis · Gordon J. Pace
Adam Wyner (Eds.)

Controlled Natural Language

5th International Workshop, CNL 2016
Aberdeen, UK, July 25–27, 2016
Proceedings

 Springer

Editors
Brian Davis
National University of Ireland
Galway
Ireland

Adam Wyner
University of Aberdeen
Aberdeen
UK

Gordon J. Pace
University of Malta
Msida
Malta

ISSN 0302-9743   ISSN 1611-3349   (electronic)
Lecture Notes in Artificial Intelligence
ISBN 978-3-319-41497-3   ISBN 978-3-319-41498-0   (eBook)
DOI 10.1007/978-3-319-41498-0

Library of Congress Control Number: 2016942903

LNCS Sublibrary: SL7 – Artificial Intelligence

Printed on acid-free paper

This Springer imprint is published by Springer Nature
The registered company is Springer International Publishing AG Switzerland

Preface

CNL 2016 was the fifth in the series of workshops on Controlled Natural Language (CNL), first organized in 2009. The 2016 edition was organized in Aberdeen, Scotland.

As with previous editions of the workshop, this year's papers cover the wide spectrum of the area of Controlled Natural Languages, ranging from human oriented to machine processable controlled languages, and from more theoretical results to interfaces, reasoning engines, and real-life applications of CNLs.

This year we invited both long and short papers to be submitted to the workshop, and we received a total of 15 submissions, of which 13 where accepted for publication based on at least two (in most cases three) reviews. In addition, the program included three invited speakers — Silvie Spreeuwenberg (LibRT), Philipp Cimiano (University of Bielefeld), and Tim Finin (University of Maryland).

We would like to thank the Program Committee for their reviews and feedback, the authors for their contributions, and the invited speakers for accepting our invitation to present their work at the workshop. Furthermore, we would also like to thank the University of Aberdeen for hosting the workshop and the workshop sponsors Digital Grammars and Insight Centre for Data Analytics, National University of Ireland Galway.

May 2016
Brian Davis
Gordon J. Pace
Adam Wyner

The original version of the preface was revised: The name of an invited speaker as well as several typos were corrected.
The erratum to this book is available at 10.1007/978-3-319-41498-0_12

Organization

CNL 2016 was organized by the Department of Computing Science of the University of Aberdeen.

Organizing Committee

Adam Wyner	University of Aberdeen, UK
Brian Davis	INSIGHT@NUI Galway, Ireland
Gordon Pace	University of Malta, Malta

Invited Speakers

Silvie Spreeuwenberg	LibRT
Philipp Cimiano	University of Bielefeld, Germany
Tim Finin	University of Maryland, USA

Program Committee

Krasimir Angelov	Chalmers University, Sweden
Paul Buitelaar	INSIGHT@NUI Galway, formerly DERI, Ireland
Rogan Creswick	Galois, USA
Brian Davis	INSIGHT@NUI Galway, formerly DERI, Ireland
Ronald Denaux	iSOCO, Spain
Esra Erdem	Sabanci University, Turkey
Sébastien Ferré	University of Rennes 1, France
Norbert E. Fuchs	University of Zurich, Switzerland
Normunds Grztis	University of Latvia
Kaarel Kaljurand	Nuance Communications, Austria
Peter Koepke	University of Bonn, Germany
Tobias Kuhn	ETH Zurich, Switzerland
John McCrae	INSIGHT@NUI Galway, formerly DERI, Ireland
Gordon Pace	University of Malta
Laurette Pretorius	University of South Africa
Aarne Ranta	University of Gothenburg, Sweden
Mike Rosner	University of Malta
Uta Schwertel	IMC Information Multimedia Communication AG, Germany
Rolf Schwitter	Macquarie University, Australia
Geoff Sutcliffe	University of Miami, USA
Irina Temnikova	Qatar Computing Research Institute, Qatar
Camilo Thorne	Free University of Bozen-Bolzano, Italy
Jeroen Van Grondelle	HU University of Applied Sciences Utrecht, The Netherlands
Adam Wyner	University of Aberdeen, UK

Sub-Reviewers

Hazem Safwat	INSIGHT@NUI Galway, formerly DERI, Ireland
Sapna Negi	INSIGHT@NUI Galway, formerly DERI, Ireland

Sponsoring Institutions

Digital Grammars Gothenburg AB
Insight Centre for Data Analytics

Contents

A Controlled Natural Language
for Tax Fraud Detection

Aaron Calafato, Christian Colombo$^{(\boxtimes)}$, and Gordon J. Pace

University of Malta, Msida, Malta
{aaron.calafato.06,christian.colombo,gordon.pace}@um.edu.mt

Abstract. Addressing tax fraud has been taken increasingly seriously, but most attempts to uncover it involve the use of human fraud experts to identify and audit suspicious cases. To identify such cases, they come up with patterns which an IT team then implements to extract matching instances. The process, starting from the communication of the patterns to the developers, the debugging of the implemented code, and the refining of the rules, results in a lengthy and error-prone iterative methodology. In this paper, we present a framework where the fraud expert is empowered to independently design tax fraud patterns through a controlled natural language implemented in GF, enabling immediate feedback reported back to the fraud expert. This allows multiple refinements of the rules until optimised, all within a timely manner. The approach has been evaluated by a number of fraud experts working with the Maltese Inland Revenue Department.

1 Introduction

Fraud is a critical aspect in any financial system, not least of which in the area of taxation. In general, fraud can be addressed by identifying rules which can either be automatically discovered or specifically defined. Much work has gone into statistical and machine learning techniques to identify rules or patterns automatically e.g. [MTVM93, TD99], however to date they have proved to be of limited effectiveness. Our focus has been to aid fraud experts in the definition of rules. This process typically consists of (i) human fraud experts identifying patterns of income and/or expenditure which might indicate fraud; (ii) these patterns are explained to an IT team, who implements the necessary code to identify matching individuals and companies from a database; (iii) based on which, the fraud expert may refine the pattern, resubmitting it each time to the IT team. This process, illustrated in Fig. 1 typically goes through many iterations until effective patterns are identified by the fraud expert to extract suspicious cases with acceptable rates of false positives and negatives. The most challenging problem is that there are many points of failure in such a workflow, and when a fraud expert gets results from a submitted pattern which are not as desired, it could be for one (or more) of various reasons, from the developer misunderstanding the informal description of the pattern to implement, to the pattern not being an effective one, or even due to a bug in the implementation.

© Springer International Publishing Switzerland 2016
B. Davis et al. (Eds.): CNL 2016, LNAI 9767, pp. 1–12, 2016.
DOI: 10.1007/978-3-319-41498-0_1

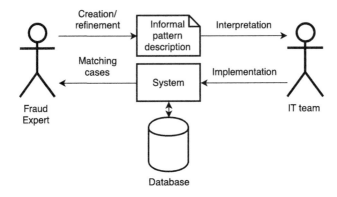

Fig. 1. Fraud detection process

Consider, for instance, the following rule which can be written by a fraud expert to identify taxpayers who declared low income for a number of years, which may be deemed to be suspicious:

Get taxpayer IDs for individuals who declared total income of less than 3000 Euro for any three years of assessment.

Some issues which may arise when applying the rule to extract reports from existing data are the following: Firstly, since the rule is expressed in plain English, it may be written imprecisely by the fraud expert, or be misinterpreted by the developer who might, for instance, interpret "total income" to refer to income from employment sources, and leave out other incomes such as pensions and bank interests. Furthermore, the manual development process is prone to coding errors, especially when dealing with multiple alterations of rules which require a number of changes.

Unexpected results due to these issues would require iterating through the process to ensure that the right reports are issued. However, these patterns identified by fraud experts are typically intelligent guesses, and it is desirable that they experiment with different rules, trying to identify ones which work better than others. Furthermore, constants such as the "3000 Euro" and the "three years" in this rule would be initial guesses intended to be refined interactively by the fraud expert. Would 3500 Euro be a better threshold in order to capture more potential fraud, or would 2500 Euro be better to get less candidate fraud cases thus affording more time to look at them more closely? For the fraud expert it is practically impossible to differentiate between the case when unexpected results are due to misinterpretation or a bug in the code, and when it is due to setting too high a threshold.

In this paper, we present an alternative approach to fraud detection, following the same methodology as currently in use, but by-passing the development process, thus reducing the points of failure. By having the fraud experts script their requests themselves, a correct compiler would allow them to focus directly on the development and refinement of fraud patterns. Since these experts are typically non-technical, the first challenge was to develop a controlled natural language (CNL) [Kuh14] for the domain of tax fraud rules, which allows them to write and execute rules directly. Another challenge is how to go from rule specifications written in the fraud CNL to executable code to process income information about entities and filter cases satisfying the rule. In order to perform this, we have used runtime monitoring [LS09] techniques to compile CNL statements compositionally into stream processors which flag matching cases.

The approach has been implemented based on the actual data submitted in the Maltese Inland Revenue Department (IRD) system using the Grammatical Framework [AR10], and the tool and language were evaluated by involving tax fraud experts. Although the test population is small (due to the small number of tax fraud experts available), indications are that the level of abstraction of our language was appropriate to enable non-technical experts to understand and write rules and execute them to obtain fraud reports. The main contribution of the paper is the investigation of the application of a controlled natural language in a real-life setting, and the use of the language by non-technical experts to define runtime monitors.

The paper is organised as follows. In Sect. 2 we describe the framework we have developed. Section 3 describes the CNL we are proposing as well as its evaluation. The translation from rules written in the CNL to executable report generators is discussed in Sect. 4, in which we also examine the efficiency of the executable rules. We discuss similar work in Sect. 5, and finally conclude in Sect. 6.

2 A Fraud Monitoring Architecture

The main challenge we have addressed is that of empowering the non-technical fraud expert to write and experiment with fraud identification rules. Furthermore, the system enables the automatic extraction of information from a large database storing data about legal entities which is regularly updated with the submission of new tax documents. We have built a solution tailored to real-life data from the Maltese Inland Revenue Department (IRD) system. The challenge can be split into two sub-problems: (i) the design of an appropriate CNL to enable the writing of rules, and (ii) how such rules can be processed on a growing database of entries to check which taxpayers match particular rules.

The former problem has been addressed by developing a CNL focused on the actual needed concepts and grammar to address tax fraud patterns and which was discussed with a fraud expert working at the IRD during meetings held prior to the design of the system. The language has been implemented in GF [AR10] and allows non-technical users to input rules avoiding syntax errors. The latter problem, that of rule processing, is that the underlying rule execution engine has to embody the semantics of the CNL, but also ensure efficient processing of data — thus the semantics of the CNL would need to be interpretable in a serial manner, allowing for incremental analysis as new data and documents are received, requiring global re-evaluation only when new rules are set up.

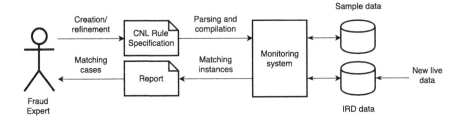

Fig. 2. Automated flow using a controlled natural language

With an ever changing dataset, another challenge for fraud experts is that of assessing the effectiveness of their new rules. We have adopted a tagged control dataset which can be used by the experts when experimenting with new rules. The limited size of the dataset ensures fast response, and the tagging (whether an entity is known to have been responsible for fraud) allows the system to report the percentage of false positives or negatives.

The resulting framework is shown in Fig. 2. This approach avoids going through a software developer every time a rule is added or modified, since the underlying concepts are hard coded into the system. The final result allows the fraud expert to refine the rules in a "what-if" manner where the feedback would indicate the accuracy of the rule. After analysing the rule on sample data, the fraud expert can target the full set of cases to be checked obtaining an accurate list of taxpayers which can be audited.

3   A CNL for the Specification of Fraud Rules

The design of the CNL was crucial to ensure that fraud experts can identify fraud cases by querying a system using known patterns. These patterns involve the querying of tax submissions including patterns between different years of assessment. The language vocabulary and grammar were based on a corpus of queries identified through consultations with fraud experts from the Maltese IRD.

Apart from a basic ontology of terms related to financial and fraud concepts (such as employee, tax payer, income, expenses, tax credits and deductions) and relations between them (e.g. an employee is a tax payer, and an employee has an income), the language had to include (i) temporal notions, to be able to refer to time points and intervals in expressions (e.g. *"the income for the current year"* or *"the average income for the previous 3 years"*); and (ii) mathematical expressions in the form of arithmetic operators and numeric comparisons (e.g. *"income + 3000 Euro"*). Furthermore, the language had to include means of referring to aggregated values (e.g. *"the income for the current year is less than the income of each of the previous three years"*).

3.1   The Language

The fraud rule language uses a number of four basic concepts: (i) taxpayer filters, (ii) time-based sets, (iii) conditions, and (iv) reports. These are then combined at the top level to obtain full rules. The basic concepts are:

Taxpayer Filters. A key element of fraud rules is the choice of a subset of taxpayers on which a check is to be made. This includes categories of taxpayers, such as *employee, pensioner* and *company*, but also filters on data properties of the taxpayers as in the greyed out phrase in the following example:

> *Load the ID, where for any year,* an employee of age more than 30 *declared an income less than 3000 Euro.*

These filters match the following grammar snippet:

$$Taxpayer ::= taxpayer \mid Individual \mid Company \mid Employer$$
$$Individual ::= IndividualType$$
$$\mid \quad IndividualType\ workAs\ Job$$
$$\mid \quad IndividualType\ age\ ComparisonOperator\ n$$
$$IndividualType ::= individual \mid director \mid employee$$
$$Company ::= CompanyType \mid CompanyType\ operatingIn\ Industry \mid \ldots$$
$$CompanyType ::= company \mid SME \mid partnership \mid \ldots$$

Time-Based Sets. Another commonly occurring sentence component is that of temporal constraints and intervals. These are used in two distinct ways in the queries (i) specifying which years the rule is to be applied on e.g. "Load ID where for any 3 years . . . ", and (ii) give context to a field e.g. ". . . the income of each year is lower than the previous". The grammar has been defined to enable the identification of sets of years which are (i) sequential e.g. *any 5 sequential years* but also relative to a point in time as in *"the previous 4 years"*; (ii) open intervals of years e.g. *"the year 2009 onwards"*; (iii) arbitrary sets e.g. *"any 2 years"*; and (iv) singleton sets e.g. *"the current year"* and *"the year 2015"*. Furthermore, these can be combined as in *"for any 3 years from the year 2009 onwards"*, which intersects two year-set selectors.

Conditions. The conditions part of the language contains the checks to be met for a case to be flagged for auditing, typically using (i) a number of fields holding values such as *total income* and *profits*; (ii) aggregation operators such as *average, total* and *minimum*; and (iii) value comparison relations such as *less than*; (iv) trends on values or their change in availability such as *increase in income* and *stopped declaring profits*. These can be combined with year-sets to constrain aggregation operators e.g. *"the average income for years after 2009 is less than 2000 Euro"*.

In general, when the values throughout the rule refer to the current year of assessment, we give the option of leaving out the time reference e.g. *"income is less than expenses"* would be considered equivalent to *"income for the current year is less than expenses for the current year"*. In order to avoid ambiguity, we only allow this omission if there are no other references to years throughout the rule.

Finally, conditions can be combined using Boolean connectives as in:

Load the ID, where for any 3 sequential years from year 2009 onwards, an employee of age more than 30 declared a total income less than 3000 Euro or declared a decrease in employment income.

Reports. Another important element of the language is the ability to identify which fields are to appear in the fraud report using regular natural language construction:

Load the ID, age and total income for the last three years, where for any 3 sequential years from year 2009 onwards, an employee of age more than 30 declared a total income less than 3000 Euro or declared a decrease in employment income.

These language elements have been combined to obtain the fraud rule specification language, which takes a reporting clause, taxpayer identification clause, year-set constraint and filtering condition. Although the general structure of a rule is rather constrained, the freedom in the individual components allows for a wide range of rules covered by the language.

The language has been built within the Grammatical Framework (GF) [Ran11]. GF is a framework designed to address the translation of multiple languages by providing a number of CNLs in different languages. Furthermore, GF provides a number of morphological paradigms which aid in making the language more readable. GF was suitable since it has allowed us to create a custom-made language and apply these morphological inflictions on the language. The custom-made language was needed since the domain was too specific to use available generic CNLs.

The grammar contains around 160 rules, with approximately 1,000 lines of code. Most of the language has been built from scratch since there are too main domain specific concepts, but reusing morphological rules from the GF framework.

GF also offers a number of authoring tools, one of which is the Minibar[1] on-line editor. This has been used to display the grammar in a more suitable format. Using a predictive parser, the tool allows only valid entries, which was vital to make sure that all the rules are grammatically correct.

3.2   CNL Evaluation

The effectiveness of the fraud rule CNL we developed has been evaluated with six fraud experts working with the Maltese IRD and two accountants working in the private sector. Although the number of persons evaluating the system is low, it is worth noting that this includes practically all fraud experts working in the field with the IRD in Malta. Full details of the evaluation can be found in [Cal16].

---

[1] http://cloud.grammaticalframework.org/gfse/.

The use of the language was evaluated during individual meetings, with each fraud expert being presented with the language in a one-hour session. The users were given a short demonstration of the language which was followed with four exercises assessing different aspects of how effective the language is: (i) *ability to read and understand the language*; (ii) *completeness of the terminology and concepts covered by the language*; (iii) *ability to write rules*.

For the first exercise, we took advantage of the fact that users were bilingual and were asked to explain a number of rules written using the CNL in Maltese, thus avoiding simple paraphrasing of the sentence. The second exercise consisted of two parts (a) they were asked to translate a number of fraud rules written in Maltese into English to assess whether the vocabulary and underlying concepts used were included in the CNL; and (b) they were asked to come up with three tax fraud rules and write them in natural language to assess whether the grammatical constructs used were allowed in the CNL. For the final test, the users were asked to pick one of the rules identified in the previous exercise and write it using the authoring tool in our CNL.

The users managed to understand almost all the rules presented to them, with only one rule which was not fully understood by two fraud experts. The rule used two instances of year selections which can lead to potential ambiguity in interpretation:

Load ID where for any 3 years, an individual declared average total chargeable income for the previous 3 years less than 2000 Euro.

Individual differences in preferences with regards to the language were minimal, and could easily be catered for by extending the vocabulary, for instance, to include the word *times* as an alternative to the mathematical symbol $*$ as preferred by one user.

When asked to identify fraud rules, and to implement one of them using the CNL, all users identified rules which were covered by the language, and were correctly expressed. However, there were cases were fraud experts wanted to refer to fields which were not available in the proof-of-concept tool developed. For the final version of the tool, this would have to be extended to the full set of fields available from the IRD database — something which can be done in a straightforward manner.

The evaluators were also asked questions regarding the use of such a language to query information as opposed to using spreadsheets (as they usually do). Only one user preferred using a spreadsheet, but adding that this preference came from years of being accustomed to using spreadsheets due to an accounting background. It was clear that while all the users became familiar with the language during the one-hour evaluation session, a longer session would be necessary in order to cover further detail and, for instance, discuss which possible rules can be written, while also giving the users more information regarding the tool. This is needed since the language can become complex when using time-based logic.

4   Monitoring Fraud Rules

One way of giving an executable semantics to the Tax Fraud CNL is to see the database of taxpayers' information as a large repository of time-stamped data, and the rules corresponding to database queries. However, such an approach requires that it is rerun whenever new data is added to the database, and when one considers that the IRD database can be rather large, this can result in hours of processing for each complex set of rules. Since the data typically arrives in temporal order, however, one can adopt a more incremental approach, processing the data as it comes in, and keeping a state to avoid recalculating things repeatedly. To contrast the approaches, consider a rule which states:

Load the ID, age and income for the current year, where for any three sequential years, a taxpayer declared an income less than 3000 Euro.

Whenever new information about some taxpayers appears, the former global approach queries the database for data from any three sequential years about a taxpayer, and processes the rest of the logic on the data collected. Unless concrete logic is added to store the fact that for some year intervals this analysis has already been done, earlier years will be reprocessed whenever new data becomes available. In contrast, if we use an incremental approach which stores whether each taxpayer has declared less than 3000 Euro in these past two years, analysis on new data can be performed efficiently by (i) checking that the threshold has not been exceeded with the data from the new year and report if the rule is satisfied, and (ii) update the state appropriately.

There has been much work in runtime monitoring [LS09] on building techniques to address such situations efficiently. It was thus decided that we adopt a standard runtime verification tool, Larva [CPS09], to process the data efficiently, using techniques from [CP13,CPA09]. Larva allows for specifications to be written in a guarded command language format, possibly structured using automata — although for the sake of encoding the semantics of our CNL, the guarded command rules sufficed.

Rules take the form of: *event | condition* ↦ *action*, indicating that whenever the event (document being submitted, data becoming available, etc.) happens, and the condition is satisfied, the action is executed. For instance, consider a rule which states that "[*some condition holds*] *for three consecutive years*". This can be encoded by introducing a variable *count* which keeps count of the number of past consecutive years (up to 3) in which the condition held. The count is initialised to zero, and the three following rules (i) increase the count when the condition is satisfied; (ii) resets the count when it is not; and (iii) reports a match when the count has reached 3^2 with the greyed out parts to be replaced by appropriate code depending on the rest of the rule:

---

[2] The semantics of the rules is such that the conditions are all checked before any actions have taken place, thus avoiding race conditions.

$$some\ condition\ holds \mid count < 3 \mapsto count + +$$
$$some\ condition\ holds \mid count = 3 \mapsto matches$$
$$\neg\ some\ condition\ holds \mid true \mapsto count = 0$$

Furthermore, Larva allows for replication of rules for different instances of the same object, thus allowing us to structure the rules above to be run for each possible taxpayer:

$$foreach\ p : TaxPayer$$
$$some\ condition\ holds \mid p.count < 3 \mapsto p.count + +$$
$$some\ condition\ holds \mid p.count = 3 \mapsto matches$$
$$\neg\ some\ condition\ holds \mid true \mapsto p.count = 0$$

In order to take appropriate action depending whether an instance matches, Larva allows for communication channels to be used between rules. Whenever a match occurs, we can send a message with the taxpayer's information on a particular channel: $matches(p)!$, which is consumed by a reporting rule. For instance, a rule starting *"Load the ID, age and income for the current year..."*, would be encoded as the follows:

$$matches(p)? \mid true \mapsto load(p.ID, p.Age, p.income(currentYear))$$

In order to implement time-based checks such as *"average income for the previous three years is less than 3000 Euro"*, the system stores information from one year to another. In order to avoid storing all the available data, the conditions need to address two aspects: (i) the respective condition to be implemented as a rule, and (ii) to store the yearly information of the field in question. These two aspects are addressed in different sets of rules, therefore the structure available in GF was crucial for this to be possible. By using a template-based approach to generate Larva code, with generic solutions such as monitors to keep track of counters and frequencies, and which are specialised to deal with the objects referred to in the CNL specification.

Using GF, we have encoded these translations in a compositional manner on the grammatical structure of the rules, encoding the monitors as a different language into which the rules are translated. Clearly, for the monitoring language, GF support for morphological inflections was not required, keeping the number of language structures in Larva smaller.

In the performance evaluation, our approach was deemed to perform well, and in fact, a sample of 53,000 records were checked in less than four seconds. This ensures that the fraud expert is given the report in a timely manner which is one of our aims. Furthermore, checking 3.2 million records, the system took around 3 min and this meant that rules can be executed on large sets of data and still retrieve feedback in a reasonable amount of time. When comparing these performance measures to traditional database queries, we found that unless carefully optimised, such queries are significantly less efficient as they would have to be reapplied globally every time new data becomes available.

Optimising such queries involves non-straightforward manipulation of the querying code to introduce indexes and tables with intermediate results, thus potentially being error prone and as such undesirable in our context.

5   Related Work

Jones et al. [JES00] have shown that with careful choice of syntax, even low-level (as opposed to natural) domain-specific languages can achieve a high-level of readability by non-technical experts — in their case, they present a combinator library to construct financial contracts defined by financial engineers. In contrast, despite the end-users' similar backgrounds in finance, in our initial meetings with fraud experts, the use of such low-level notation was frowned upon, which led us to go for a more natural, albeit controlled, syntax.

For similar reasons, we avoided the use of a template-based approach e.g. [PSE98], in order to allow for a more granular and compositional grammar. The approach used in our work combines the grammar-based and template-based approaches by presenting a high-level template to the users to make it more understandable, whilst still using a pool of underlying core concepts.

The use of controlled natural languages as a means of providing a flexible input format for non-technical experts to be able to express instructions is rather widespread e.g. [CGP15]. The natural aspect of the language, especially if used with an appropriate user interface supporting syntactically correct-by-design input, allows for end-user development within specific domains. GF itself has been used for various such case studies e.g. [DRE13,RED10]. What distinguishes our approach from most other similar work is the semantic interpretation of the language, and the use of a runtime monitoring approach to allow for stream-based data analysis derived from natural language queries. For instance, Dannélls et al. [DRE13], build a CNL tackling museum system queries. As in our case, theirs is focused on a particular domain, that of identifying paintings in a system. In contrast, however, our language is more controlled and technical and less natural than theirs, which was required to be able to give a semantic interpretation of the terms into a stream processing monitor.

We have already explored the use of monitoring techniques as a backend for a CNL in [CGP15]. However, the fraud language we present in this paper is substantially more extensive and expressive, even if backend techniques are similar.

6   Conclusions

In this paper, we have presented a controlled-natural-language-driven framework aimed at supporting fraud experts to be able to, in an autonomous manner, explore different fraud rules and apply them to a live set of taxpayer data. The backend of the framework has been developed as an incremental monitor, enabling sufficiently efficient analysis of large datasets — experiments show that running a number of rules on the data from 53,000 documents takes less than four seconds, an improvement on simple database queries. This limited dataset, however, is used just to enable fraud experts to assess their rules, before running them on the whole dataset with over 3.2 million documents and which were processed in approximately three minutes.

The CNL we have developed has been evaluated by a number of fraud experts currently working on real-life tax fraud detection, which showed that the language was effective both in enabling them to write rules, but also to correctly understand and interpret rules written by others. The language currently contains the basic concepts of numeric and monetary values in order to enable fraud experts to use it, however, we plan to extend the language to (i) the full set of fields found in the taxation documents; and (ii) include richer operations e.g. extend the predicate "increase in income" to be able to access a finer grained "percentage increase in income".

One interesting aspect in our use of GF is the multi-lingual support it provides which can be harnessed to present the CNL in multiple languages while keeping the same monitoring system. Since many taxation concepts are common across countries, we envisage that this should be feasible with only minor localisation issues. Although we have not addressed this in our work, we have instead treated our compilation into monitors as a GF translation. Although our approach is very constrained, it might be worth investigating further the use of translation support for compilation into executable code as a means of semantic analysis or inspection.

References

[AR10]   Angelov, K., Ranta, A.: Implementing controlled languages in GF. In: Fuchs, N.E. (ed.) CNL 2009. LNCS, vol. 5972, pp. 82–101. Springer, Heidelberg (2010)

[Cal16] Calafato, A.: A domain specific property language for fraud detection to support agile specification development. Master's thesis, University of Malta (2016)

[CGP15] Colombo, C., Grech, J.-P., Pace, G.J.: A controlled natural language for business intelligence monitoring. In: Biemann, C., Handschuh, S., Freitas, A., Meziane, F., Métais, E. (eds.) NLDB 2015. LNCS, vol. 9103, pp. 300–306. Springer, Heidelberg (2015)

[CP13]   Colombo, C., Pace, G.J.: Fast-forward runtime monitoring — an industrial case study. In: Qadeer, S., Tasiran, S. (eds.) RV 2012. LNCS, vol. 7687, pp. 214–228. Springer, Heidelberg (2013)

[CPA09] Colombo, C., Pace, G.J., Abela, P.: Offline runtime verification with real-time properties: a case study. In: Proceedings of WICT 2009 (2009)

[CPS09] Colombo, C., Pace, G.J., Schneider, G.: Larva – safer monitoring of real-time java programs (tool paper). In: Seventh IEEE International Conference on Software Engineering and Formal Methods (SEFM), pp. 33–37. IEEE Computer Society, November 2009

[DRE13] Dannélls, D., Ranta, A., Enache, R.: Multilingual grammar for museum object descriptions. In: Frontiers of Multilingual Grammar Development, p. 99 (2013)

[JES00] Jones, S.P., Eber, J.M., Seward, J.: Composing contracts: an adventure in financial engineering (functional pearl). In: ICFP 2000: Proceedings of the Fifth ACM SIGPLAN International Conference on Functional programming, pp. 280–292. ACM, New York (2000)

[Kuh14] Kuhn, T.: A survey and classification of controlled natural languages. Comput. Linguist. **40**(1), 121–170 (2014)

[LS09] Leucker, M., Schallhart, C.: A brief account of runtime verification. J. Log. Algebr. Program. **78**(5), 293–303 (2009)

[MTVM93] Maes, S., Tuyls, K., Vanschoenwinkel, B., Manderick, B.: Credit card fraud detection using bayesian, neural networks. In: Maciunas, R.J. (ed.) Interactive Image-guided Neurosurgery, pp. 261–270. American Association Neurological Surgeons, Rolling Meadows (1993)

[PSE98] Power, R., Scott, D., Evans, R.: What you see is what you meant: direct knowledge editing with natural language feedback. In: ECAI, pp. 677–681 (1998)

[Ran11] Ranta, A.: Grammatical Framework: Programming with Multilingual Grammars. Center for the Study of Language and Information/SRI (2011)

[RED10] Ranta, A., Enache, R., Détrez, G.: Controlled language for everyday use: the MOLTO phrasebook. In: Rosner, M., Fuchs, N.E. (eds.) CNL 2010. LNCS, vol. 7175, pp. 115–136. Springer, Heidelberg (2012)

[TD99] Tax, D.M.J., Duin, R.P.W.: Data domain description using support vectors. In: Proceedings of the European Symposium on Artificial Neural Networks, pp. 251–256 (1999)

Reasoning in Attempto Controlled English: Non-monotonicity

Norbert E. Fuchs[✉]

Department of Informatics and Institute of Computational Linguistics,
University of Zurich, Zurich, Switzerland
fuchs@ifi.uzh.ch
http://attempto.ifi.uzh.ch/

Abstract. RACE is a first-order reasoner with equality for Attempto Controlled English (ACE) that can show the consistency of a set of ACE axioms and deduce ACE theorems and ACE queries from ACE axioms. This paper presents various forms of non-monotonic reasoning.

Keywords: Controlled natural language · Attempto Controlled English · ACE · RACE · Monotonic reasoning · Non-monotonic reasoning · Abduction

1 Introduction

Attempto Controlled English (ACE)[1] is a logic-based knowledge representation language that uses the syntax of a subset of English. The Attempto Reasoner RACE[2] – one of several reasoners available for ACE[3] – allows users to show the consistency of an ACE text, to deduce one ACE text from another one, and to answer ACE queries from an ACE text. More about ACE and RACE is found in the relevant documentation[4].

A previous system description [1] of RACE detailed its structure, its functionality, its implementation and its user interfaces. The main part of [1] covered reasoning with what could be called the first-order subset of ACE, that is all ACE constructs – including alethic modality – that can be directly or indirectly mapped to first-order formulas. Furthermore, [1] described summation, a form of second-order reasoning combining the results of first-order proofs.

A closer look reveals that all reasoning examples found in [1] have one feature in common, namely that they rely on monotonic logic, i.e. adding axioms will not invalidate deductions from the original axioms.

This paper extends RACE's reasoning to non-monotonic logic[5] and presents solutions to some of its many facets. Please note that the paper is a further system description focussing on RACE's implementation. Thus you will not learn much about

---

[1] http://attempto.ifi.uzh.ch/.
[2] http://attempto.ifi.uzh.ch/race/.
[3] http://attempto.ifi.uzh.ch/site/resources/.
[4] http://attempto.ifi.uzh.ch/site/docs/.
[5] https://en.wikipedia.org/wiki/Non-monotonic_logic.

© Springer International Publishing Switzerland 2016
B. Davis et al. (Eds.): CNL 2016, LNAI 9767, pp. 13–24, 2016.
DOI: 10.1007/978-3-319-41498-0_2

non-monotonic reasoning in general beyond what is needed to explain the selected topics and examples.

The rest of this paper is organised as follows. In Sect. 2, I recall general features of RACE. Section 3 recounts summation and shows that it has both monotonic and non-monotonic aspects. In Sect. 4, I focus on variations of non-monotonic reasoning. Section 5 presents a form of abduction. Section 6 concludes with a summary of the presented solutions and with a discussion of their strengths and limitations, specifically addressing the question of their suitability.

2 General Features of RACE

For the convenience of the reader and to make this paper self-contained the material of this section is partially copied from [1].

RACE has the following general features:

- RACE offers consistency checking, textual entailment and query answering of ACE texts.
- For inconsistent ACE axioms RACE will list all minimal subsets of the ACE axioms that lead to inconsistency. If ACE axioms entail ACE theorems, respectively ACE queries, RACE will list all minimal subsets of the ACE axioms that entail the theorems, respectively queries.
- RACE is loosely based on the model-generator Satchmo [2], but offers much additional functionality.
- RACE is implemented as a set of Prolog programs that can be used locally. Furthermore, RACE can be accessed remotely via its web-client[6] or via its web-service[7].
- RACE uses about 50 auxiliary axioms to represent domain-independent general knowledge. These auxiliary axioms are not expressed in ACE, but in Prolog.
- Using the entailment A ⊢ T of ACE theorems T from ACE axiom A as an example, here is a sketch RACE's proof procedure.

In a first step, the ACE axioms A and the ACE theorems T are separately translated by the Attempto Parsing Engine (APE)[8] into semantic representations called discourse representation structures DRS_A, respectively DRS_T [3].

In a second step, (DRS_A & ¬ DRS_T) is translated into a set of clauses clause (Body, Head, Index) where Body is true or a conjunction of logical atoms and Head is fail or a disjunction of logical atoms. Body implies Head. Index contains the number of the ACE axiom, respectively ACE theorem, from which the clause was derived. Negation is expressed as implication to fail.

In a third step, the clauses are executed bottom-up by forward-reasoning. If the Body of a clause is true or can be proved from the data base then Head is asserted to

---

[6] http://attempto.ifi.uzh.ch/race/.

[7] http://attempto.ifi.uzh.ch/ws/race/racews.perl.

[8] http://attempto.ifi.uzh.ch/site/resources/.

the database together with a list that contains Index and the sentence indices of all clauses that were used to prove Body. This amounts to building a proof-tree labelled with lists of indices. Range restriction[9] ensures that all atoms added to the data base are ground. If Head is fail then the respective branch of the proof-tree is considered closed, the indices of its leaf are stored and its nodes are removed from the data base.

Forward-reasoning ends when no (further) clause can be executed. The result is either a set of ground atoms in the data base constituting a minimal Herbrand model of the clauses – indicating that the clauses are consistent, the ACE axioms do not entail the ACE theorems – or an empty data base and a set of closed branches. If all branches of the proof-tree are closed – indicating a succeeding proof– then the indices of their leaves are combined and mapped to their respective ACE axioms that constitute a minimal subset of the ACE axioms needed to entail the ACE theorems. Since RACE checks all possibilities to prove the bodies of the clauses, it can generate more than one proof.

3   Monotonic and Non-monotonic Summation

If ACE axioms entail ACE theorems RACE will list all minimal subsets of the axioms that entail the theorems. In other words, RACE will generate more than one proof of a theorem if the axioms allow for this. Here is a simple example shown as a screen-shot of the output window of RACE's web-client.

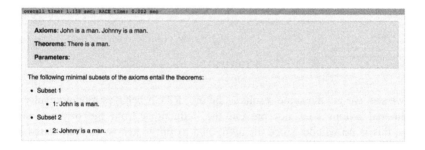

The example relies on RACE using the unique name assumption, i.e. the men *John* and *Johnny* are by default distinct. Thus there are two proofs for the theorem *There is a man.*

RACE provides also summation, i.e. second-order aggregation over first-order proofs. Using the same axioms and the theorem *There are two men.* we get.

---

[9] A clause is called range-restricted if all variables of its head occur already in its body. Clauses derived from discourse representation structures are by default range-restricted.

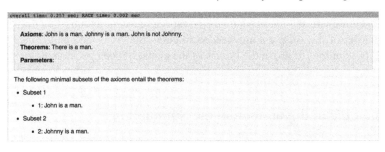

Now let's add the axiom *John is not Johnny.* and try both proofs again.

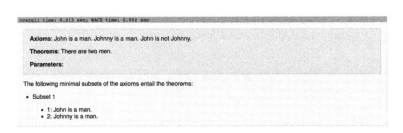

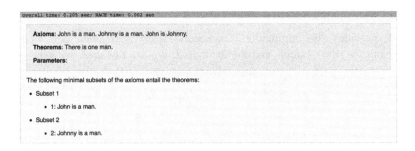

As we see, we get the same results as before. RACE behaves monotonically since the additional axiom does not prevent the deductions from the original axioms. Actually, this is no wonder since the additional axiom is just an explicit statement of the unique name assumption.

Alternatively, let's add the axiom *John is Johnny.* and try both proofs again.

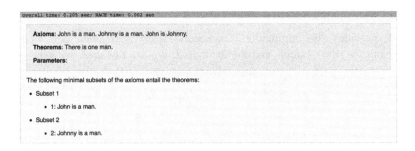

```
overall time: 0.262 sec; RACE time: 0.003 sec
```

Axioms: John is a man. Johnny is a man. John is Johnny.

Theorems: There are two men.

Parameters:

There is 1 message.

Importance	Type	Sentence	Problem	Description/Suggestion
warning	aggregation		The names "Johnny" and "John" refer to the same object of the same class. Thus there will be less objects of this class than perhaps expected.	

Theorems do not follow from axioms.

The following parts of the theorems/query could not be proved:

- countable common noun: (at least 2) man
- no abducted axioms

While the theorem *There is a man.* can be reproduced as before, the theorem *There are two men.* cannot since the unique name assumption is explicitly overridden, and we have only one man that happens to carry two names. Thus in this case RACE behaves non-monotonically since the additional axiom prevents a deduction from the original axioms.

To sum up, RACE exhibits both monotonic and non-monotonic behaviour for summation. While this behaviour was unintended, it motivated me to extend RACE by non-monotonic reasoning in general.

4 Non-monotonic Reasoning

Reasoning is called monotonic when derived conclusions are not invalidated by added premises. However, there are also highly relevant forms of reasoning – called non-monotonic – that derive tentative conclusions from assumed premises. Both the assumed premises and the tentative conclusions may have to be withdrawn or replaced in the light of further evidence. Non-monotonic reasoning plays an important role in everyday thinking and argumentation, in semiformal settings like medical diagnosis, or in formal domains like expert systems and physics.

Since the 1980s various formal frameworks have been developed for non-monotonic reasoning [4]. In the last years the focus has been on logic-based approaches like Answer Set Programming (ASP)[10]. Given RACE's foundation in first-order logic, it is not surprising that also its approach to non-monotonic reasoning is logic-based.

Of the many different forms of non-monotonic reasoning[11] I will consider two:

- Contradictory premises where one or the other premise may have to be withdrawn or modified to remove the contradiction; this case is traditionally not subsumed under the label "non-monotonic reasoning", but shows many similarities.
- Default reasoning where conclusions depend on default premises that express knowledge that is typically, but not necessarily always, true. Default reasoning comes in many variations some of which will be addressed.

---

[10] http://www.kr.tuwien.ac.at/staff/tkren/pub/2009/rw2009-asp.pdf.

[11] http://plato.stanford.edu/entries/logic-nonmonotonic/.

Contradictory Axioms. The next example was specifically chosen to show that contradictory axioms are not necessarily so easily perceived that the user becomes aware of the contradiction. RACE, however, will detect it and notify the user.

RACE handles alethic modality (necessity, possibility) by mapping modal language constructs to standard first-order logic constructs[12]. The following example makes use of the fact that the modal statement *John must sleep.* entails both the modal statement *John can sleep.* and the non-modal statement *John sleeps.* Thus the two axioms are contradictory. Notice that – as if RACE would make an effort to give the user as much information as possible – the theorem is nevertheless derived from the first axiom. More importantly, there is a warning message and the user must decide how to remove the contradiction.

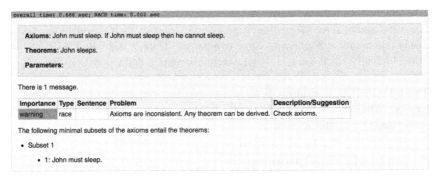

Worse, contradictory axioms allow users to derive a theorem – e.g. *John wakes.* – that does not follow at all from the axioms though it seems related to the subject matter. RACE cannot prevent its users from constructing cases like this one, but alerts them by issuing a warning message, and inviting them to change the axioms.

```
overall time: 0.206 sec; RACE time: 0.002 sec
```

Axioms: John must sleep. If John must sleep then he cannot sleep.

Theorems: John wakes.

Parameters:

There is 1 message.

Importance	Type	Sentence	Problem	Description/Suggestion
warning	race		Axioms are inconsistent. Any theorem can be derived.	Check axioms.

The following minimal subsets of the axioms entail the theorems:

- Subset 1
    - 1: John must sleep.
    - 2: If John must sleep then he can not sleep.

---

[12] https://en.wikipedia.org/wiki/Standard_translation.

Default Reasoning: Defeasible Information. John is checking a train time-table for a specific train. Not finding this train he concludes that it does not exist. John's reasoning is based on the so-called *closed-world assumption*[13] that states that the information available is assumed complete, that everything that is true is known to be true, and that everything that is not known to be true is considered false. The *closed-world assumption* leads to defeasible conclusions, i.e. conclusions that may have to be withdrawn in the light of new evidence. When John detects that he by mistake checked an outdated time-table and that the actual time-table does contain the train he is looking for, he has to revise his previous conclusion.

In addition to logical negation ACE provides language constructs (... *does/do/is/are not provably..., it is not provable that...*) that stand for negation as failure (NAF)[14]. Using these language constructs together with logical negation I implemented the *closed-world assumption* in RACE. RACE allows negation as failure constructs only in the preconditions of ACE's if-then statements which leads to entries `naf(NAF)` in the bodies of RACE's clauses. `NAF` is a list of logical atoms that RACE tries to prove as described in Sect. 2. If the proof of `NAF` succeeds then `naf(NAF)` fails, and vice versa.

Here is the train example for the initial situation when John searches the outdated time-table, and – not finding the specific train – concludes that it does not exist.

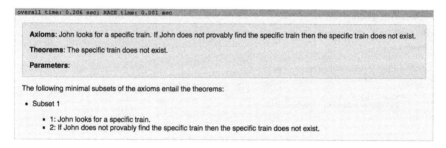

If John looks for and finds the specific train in the actual time-table, he no longer deduces that the train does not exist. RACE also reproduces this behaviour.

```
overall time: 0.209 sec; RACE time: 0.003 sec

Axioms: John looks for a specific train and finds it. If John does not provably find a specific train then the specific train does not exist.

Theorems: A specific train does not exist.

Parameters:

Theorems do not follow from axioms.
The following parts of the theorems/query could not be proved:
  • no abducted axioms
```

[13] https://en.wikipedia.org/wiki/Closed-world_assumption.
[14] https://en.wikipedia.org/wiki/Negation_as_failure.

Default Reasoning: Working With Exceptions. An often-cited example of default reasoning concerns the fact that birds typically fly, but that there are exceptions like penguins that are birds, but do not fly. Thus the statement *All birds fly*. does not correctly represent reality. Replacing this statement by *All birds fly with the exception of penguins, ostriches, kiwis, dead birds, young birds, wounded birds, ...* obviously is not practicable for automatic reasoning since the list of non-flying birds would have to be updated again and again.

As in the previous case negation as failure in combination with logical negation offers a convenient way to express typicality and its exceptions.

We can represent the bird example in two ways, once with a rule that birds fly if there is no evidence to the contrary, and once with a rule that birds do not fly if there is no evidence that in fact they do.

The next two figures show these two cases. Notice that in the first case the negation as failure and the logical negation occur in the precondition of the if-then statement that expresses the typicality plus exception, while in the second case the negation as failure occurs in the precondition and the logical negation in the consequence. Which formulation to choose depends among other considerations on the theorem to be proved.

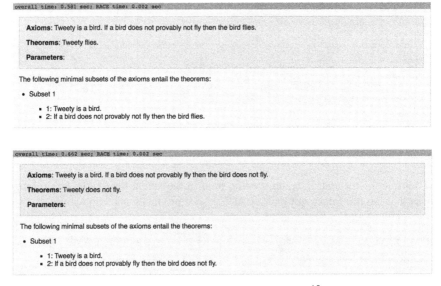

Default Reasoning: Frame Problems. The frame problem[15] of artificial intelligence emerges when one parameter of a complex situation is changed and the question arises how to represent that all other parameters of the situation that are not dependent on the changed parameter stay unchanged. Of the many solutions to the frame problem those that are based on default logic[16] and on answer set programming (ASP)[17] seem to be

---

[15] https://en.wikipedia.org/wiki/Frame_problem.

[16] https://en.wikipedia.org/wiki/Frame_problem#Default_logic_solution.

[17] https://en.wikipedia.org/wiki/Frame_problem#Answer_set_programming_solution.

the simplest since they directly express the common sense law of inertia, also called Leibniz law, that everything can be assumed to remain in the state in which it is.

Taking the ASP formulation of the frame axiom – *if r(X) is true at time T, and it can be assumed that r(X) remains true at time T + 1, then we can conclude that r(X) remains true* – as a guidance here is an example how the frame problem can essentially be solved by RACE.

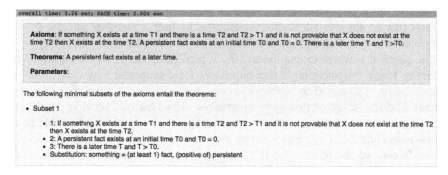

The first axiom – like ASP relying on a combination of logical negation and negation as failure – expresses the persistence of situation parameters not affected by the change of another parameter. The second and third axiom describe an elementary situation.

Default Reasoning: Nixon Diamond. Default reasoning can also lead to an impasse. In the problem called Nixon diamond default assumptions lead to mutually inconsistent conclusions that require additional reasoning steps. The situation is as follows: Nixon is both a quaker and a republican. Quakers tend to be pacificists, while republicans usually do not. From these premises one can deduce that Nixon is both a pacifist and no pacifist.

Please note that in the following screen-shot the word *quaker* is prefixed by *n:* to identify it as noun since *quaker* is not found in RACE's lexicon.

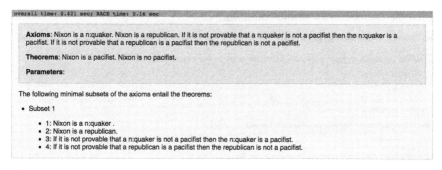

To get out of this impasse there are several possibilities – all of which could in principle be implemented in RACE, but are not. First, one can assume either a skeptical attitude – the conflicting conclusions must not be drawn – or a credulous attitude – the conflicting conclusions are derived, but must be resolved in one way or other in the light of further evidence. Second, like [5] one can introduce priorities for the two default rules so that only one conclusion is derived.

5 Abduction

Abduction – another form of non-monotonic reasoning – strives to find a likely explanation for a phenomenon. In terms of reasoning this means extending given premises that do not give the desired conclusions by further premises selected from a background theory, possibly subject to constraints like simplicity or plausibility.

The question is where to find the additional premises. One answer is suggested by Abductive Logic Programming[18] that introduces logic programs with clauses whose bodies consists of so-called abducible predicates that are only partially defined and that possibly underlie a set of first-order constraints. The idea of abductive logic programming is to extend the given logic program by definitions of the abducible predicates – respecting the constraints – so that the extended program generates the desired answer. In essence, the information that leads to abducted premises is contained in the bodies of the given clauses.

This approach inspired RACE's strategy for a form of abduction:

- axioms contain if-then statements whose preconditions are incompletely defined; possibly there are some constraints for these preconditions
- theorems are matched against the conclusions of the if-then statements; matching may have to extend to all given axioms if noun phrases occur in the conclusions only as anaphoric references, but are actually defined elsewhere
- if matching succeeds then the precondition of the respective if-then statement is added to the axioms respecting the constraints and avoiding inconsistency and duplication

Here is a simple example[19]: If your lawn is wet then it either rained or you had switched on the sprinkler. If the sky is clear you must exclude rain as the cause, so that the sprinkler is the cause.

First the example without the constraint.

overall time: 0.213 sec; RACE time: 0.013 sec

Axioms: There is a lawn. If there is some rain then the lawn is wet. If the sprinkler runs then the lawn is wet.

Theorems: The lawn is wet.

Parameters:

Theorems do not follow from axioms.

The following parts of the theorems/query could not be proved:

- adjective: wet
- copula: is/are
- abducted axiom to prove the theorem: There is a sprinkler X1. The sprinkler X1 runs.
- abducted axiom to prove the theorem: There is some rain.

---

[18] https://en.wikipedia.org/wiki/Abductive_logic_programming.

[19] https://en.wikipedia.org/wiki/Abductive_logic_programming#Example_1.

We get two sets of abducted axioms that each – together with the original axioms – prove the theorem.

Now we add the constraint that it cannot rain from a clear sky and the fact that the sky is clear.

```
overall time: 0.205 sec; RACE time: 0.013 sec
```

Axioms: There is a lawn. If there is some rain then the lawn is wet. If the sprinkler runs then the lawn is wet. It is false that the sky is clear and that there is some rain. The sky is clear.

Theorems: The lawn is wet.

Parameters:

Theorems do not follow from axioms.

The following parts of the theorems/query could not be proved:

- adjective: wet
- copula: is/are
- abducted axiom to prove the theorem: There is a sprinkler X1. The sprinkler X1 runs.

As expected we get only one set of abducted axioms that together with the original axioms prove the theorem.

6 Conclusions

I extended the Attempto Reasoner RACE by non-monotonic reasoning using small examples that focus on the respective issues.

As described in [1] RACE relies on auxiliary axioms that add domain-independent general knowledge to the domain-specific knowledge of the given axioms. Since these auxiliary axioms are coded in Prolog that has the power of the Turing machine, RACE could in principle deduce any conclusion from the axioms. As a consequence – in addition to the usual questions of correctness and completeness of a theorem prover – a further question arises, namely what RACE should actually deduce. The answer to this question depends on the domain investigated and on the expectations and intuitions of the users, and – as my experience has shown – may be highly debatable.

In non-monotonic reasoning a similar question of suitability arises, this time related to the reasoning methods I chose. While default reasoning based on ACE's language constructs negation and negation as failure seems quite general and powerful, the same cannot yet be said for RACE's strategy for abduction inspired by abductive logic programming. Thus questions arise, for example which of my methods is dictated by the example problems, and which is powerful and general enough to be applied to other types of problems. In other words, the methods and implementations of non-monotonic reasoning that I presented in this paper need further investigations.

RACE now covers all language constructs of Attempto Controlled English with the exception of those that have no direct logical representations – imperative sentences and the modal operators for recommendation (*should*) and admissibility (*may*) – and the operations on lists, sets and strings that – not posing any problems as far as reasoning is concerned – will be implemented in RACE at some other time.

Acknowledgements. I would like to thank the three anonymous reviewers of the first version of this paper for their constructive comments. Many thanks go to the Department of Informatics and the Institute of Computational Linguistics, University of Zurich, for their hospitality.

References

1. Fuchs, N.E.: First-order reasoning for Attempto Controlled English. In: Rosner, M., Fuchs, N.E. (eds.) CNL 2010. LNCS, vol. 7175, pp. 73–94. Springer, Heidelberg (2012)
2. Manthey, R., Bry, F.: SATCHMO: a theorem prover implemented in prolog. In: Overbeek, R., Lusk, E`. (eds.) CADE 1988. LNCS, vol. 310, pp. 415–434. Springer, Heidelberg (1988)
3. Bos, J.: Computational semantics in discourse: underspecification, resolution, and inference. J. Logic Lang. Inf. **13**, 139–157 (2004). Fuchs, N.E., Kaljurand, K., Kuhn, T.: Discourse Representation Structures for ACE 6.6, Technical Report ifi-2010.0010, Department of Informatics, University of Zurich (2010)
4. Genesereth, M.R., Nilsson, N.J.: Logical Foundations of Artificial Intelligence. Morgan Kaufmann Publishers, San Mateo (1987)
5. Grosof, B.N.: Courteous logic programs: prioritized conflict handling for rules. IBM Research Report RC 20836. IBM T. J. Watson Research Center (1997)

Bootstrapping a Runyankore
CNL from an isiZulu CNL

Joan Byamugisha$^{(\boxtimes)}$, C. Maria Keet, and Brian DeRenzi

Department of Computer Science, University of Cape Town, Cape Town, South Africa
{jbyamugisha,mkeet,bderenzi}@cs.uct.ac.za

Abstract. Runyankore is one of the top five languages spoken in Uganda. It is a Bantu language, thus it possesses the characteristic agglutinative structure, known to be challenging for the development of computational resources. It is also computationally under-resourced, which compounds the problem further. Given the recent progress in verbalization (writing the semantics expressed in axioms as CNL) of most constructors in the Description Logic $\mathcal{ALC}$ in isiZulu, we take a bootstrapping approach to verbalization of similar constructors in Runyankore. The key variables affecting verbalization in isiZulu indeed also hold for Runyankore, allowing us to build on existing background theory. We present verbalization patterns for most $\mathcal{ALC}$ constructors, also covering the 'hasX' role naming. Evaluation of text generated with 18 non-linguists found a clear preference for verbalization in the singular for subsumption (as with isiZulu), existential quantification, and negation in the context of subsumption; but the plural form of verb negation.

Keywords: Bootstrapping · Runyankore · isiZulu · Verbalization · $\mathcal{ALC}$ constructors · CNL

1 Introduction

Runyankore is one of the top five most widely spoken languages in Uganda [2,18,20]. Wider internet access in Africa has led to the expansion of technology localization for indigenous languages. Large technology companies, such as Google and Mozilla, have made efforts to provide localized versions of their software. For Uganda, the Google search engine is available in Ekinyarwanda, Kiswahili, Luganda, Luo, and Runyakitara.

For any language, the development of computational components is resource-intensive [4]. However, when dealing with a very under-resourced language like Runyankore, it is justifiable to tailor what has already been done in a similar language, as a means of reducing development time and effort [4,7]. One way of doing this is through the bootstrapping approach. Specifically, and similarly to [4], this means using the existing isiZulu verbalization patterns to generate sentences from axioms in a Description Logic (DL) as the starting point from which Runyankore ones can be tailored.

© Springer International Publishing Switzerland 2016
B. Davis et al. (Eds.): CNL 2016, LNAI 9767, pp. 25–36, 2016.
DOI: 10.1007/978-3-319-41498-0_3

Bootstrapping has been applied to develop CNLs using templates [11] and the Grammatical Framework (GF) [1]. However, the template approach is inapplicable to languages with an agglutinating morphology [13,14], and Runyankore is too under-resourced for GF, which requires a large resource grammar library; hence the need for bootstrapping for CNLs of agglutinated languages based on a grammar engine, which has not yet been done. The benefits of the bootstrapping approach for the development of language resources has been documented [4,7,12], most important of which is a reduction in development time and effort without sacrificing accuracy [4,7].

We therefore seek to find out whether: (1) verbalization in Runyankore is affected by the same variables as isiZulu, namely the noun class of the name of the concept, the category of the concept, whether the concept is atomic or an expression, the quantifier used in the axiom, and the position of the concept in the axiom; and if so, (2) the existing isiZulu verbalization patterns can be tailored to Runyankore grammar to generate correct Runyankore text.

We define a Controlled Natural Language (CNL) as a constructed language based on a particular natural language, with a restricted lexicon, syntax, and semantics, but which still maintains its natural properties [15], and verbalization as the process of writing the semantics expressed in axioms as a CNL [11]. As a substantial amount of rules are necessary for the Runyankore CNL, it effectively blurs the line with NLG in the back-end. The Attributive Concept Language with Complements ($\mathcal{ALC}$) was used because of its expressiveness [3]. Here, we present how Runyankore verbalizations of subsumption ('is a'), conjunction ('and'), negation ('not'), existential quantification ('at least one'), and universal quantification ('for all'/'each') were tailored from isiZulu. The evaluation with 18 non-linguists showed preference for the singular form in the verbalisation for most constructors.

The paper is structured as follows: Sect. 2 introduces the basics of Runyankore; Sect. 3 presents related work in bootstrapping for language resource development; Sect. 4 illustrates how Runyankore verbalization patterns have been tailored from those of isiZulu; the evaluation is presented in Sect. 5; Sect. 6 discusses the implications of this work; and we conclude in Sect. 7.

2   Basics of Runyankore

Runyankore is a Bantu language spoken in the south-western part of Uganda by over two million people and is one of the top five most widely spoken languages in Uganda [2,18,20]. Runyankore, like other Bantu languages, is a highly agglutinative language [2,18]—a word can be composed of over five constituents [6]. As the following example demonstrates, word formation involves the addition of affixes to a base word, where each affix carries meaning such as tense and aspect [18] (AU: augment, PRE: prefix, NC: noun class, CONT: continuous tense marker, SC: subject concord, FV: final vowel):

Abaana nibazaana   A-ba-aana ni-ba-zaan-a
'The children are playing'   AU-PRE$_{NC2}$-child CONT-SC$_{NC2}$-play-FV

There are several similarities between the structure of isiZulu and Runyankore, which make it seem feasible to tailor isiZulu verbalization patterns to Runyankore grammar. In both languages, the verbal morphology is very complex, with five different tenses in isiZulu [13] and fourteen in Runyankore [20]. Each noun is associated with one of several noun classes [6,14]. The noun class determines the affixes of the nouns belonging to it, and subsequently agreement markers on the associated lexical categories such as adjectives and verbs [6,13]. Noun class prefixes are coupled as singular/plural pairs [14]. Nouns comprise of two formatives, the prefix and the stem [14], where prefixes express number and are used to determine the class to which a particular noun belongs [14]. Table 1 shows the noun class system for isiZulu and Runyankore, using Meinhof's 1948 noun class system, which is a standard for defining noun classes among linguists, and thus facilitates cross-language comparisons and use. There are, however, several differences that make the direct reuse of the isiZulu verbalization patterns impossible (e.g., vocabulary and grammatical differences).

Table 1. IsiZulu and Runyankore noun classes, with their respective standard list of prefixes; NC: Noun class, AU: augment, PRE: prefix, n/a: class is not used.

NC	isiZulu		Runyankore		NC	isiZulu		Runyankore	
	AU	PRE	AU	PRE		AU	PRE	AU	PRE
1	u-	m(u)-	o-	mu-	9	i(n)-	-	e-	n-, m-
2	a-	ba-	a	ba-	10	i-	zi(n)-	e-	n-
1a	u-	-	n/a		11	u-	(lu)-	o-	ru-
2a	o-	-	n/a		(10)	i-	zi(n)-	n/a	
3a	u-	-	n/a		12	n/a		a-	ka-
(2a)	o-	-	n/a		13	n/a		o-	tu-
3	u-	m(u)-	o-	mu-	14	u-	bu-	o-	bu-
4	i-	mi-	e-	mi-	15	u-	ku-	o-	ku-
5	i-	(li)-	e-	i-, ri-	6	n/a		a-	ma-
6	a-	ma-	a-	ma-	16	n/a		a-	ha-
7	i-	si-	e-	ki-	17	-	ku-	-	ku-
8	i-	zi-	e-	bi-	18	n/a		o-	mu-
9a	i-	-	n/a		20	n/a		o-	gu-
(6)	a-	ma-	n/a		21	n/a		a-	ga-

3   Related Work

Davel and Barnard [7] applied the bootstrapping approach to extend a German pronunciation dictionary, using automatically extracted rules to generate additional word/pronunciation pairs, which were then used to extract better rules [7]. They managed to reduce development time to less than a quarter of that required for manual development and still maintain a high level of accuracy [7].

Bootstrapping has also been used in the development of CNLs [1,11]. Jarrar et al. [11] used templates to develop multilingual verbalizations of logical theories. The initial verbalization template file would be tailored to a grammatically related language by varying the text in the text tags and their position, to reflect the language structure of the target language [11]. Through the bootstrapping approach, e.g., a German verbalization template was tailored from the Dutch template [11]. Angelov and Ranta [1] instead used GF to develop their CNL first in English, and this was then ported to Finish, French, German, Italian, and Swedish. Bootstrapping reduced development time from four days for English to a matter of hours for each of the languages [1].

For Bantu languages, however, the biggest drawback to developing computational language resources is their complex agglutinating morphology [4,13,14]. Bosch et al. [4] considerably reduced development time by applying an experimental bootstrapping approach in developing morphological analyzers for isiXhosa, Swati, and Ndebele based on the existing one for isiZulu [4]. Starting with the isiZulu morphological analyzer, they made the following language-specific modifications for each language: word roots lexicon, grammatical morpheme lexicon, as well as the language appropriate morphophonological rules [4]. The reduction in development time was from over 3,000 h for the initial isiZulu morphological analyzer to a total of about 300 h for all three new ones, with a further improvement using language-specific resources and rules from an average accuracy of 71.3 % to 95.6 % [4].

In summary, it has been shown that the bootstrapping approach has so far been used to develop a pronunciation dictionary, morphological analyzer, and generate language using templates and GF. As explained in Sect. 1, templates cannot be applied to an agglutinative language like Runyankore, and GF requires a wide coverage grammar specification which currently does not exist in Runyankore, and will require a lot of time and effort to develop.

4   Comparing Verbalization in IsiZulu and Runyankore

In the bootstrapping approach, machine-learning analyses are corrected by a human "trainer" and the corrections are then further used to update the system's rules, and the cycle continues until the system achieves a satisfactory level of accuracy [12] (as was the case in [4,7]). We apply the same concept but perform a manual analysis instead of machine learning. Instead of developing Runyankore verbalization patterns from scratch, isiZulu verbalization patterns are manually analyzed and tailored to the Runyankore grammar and lexicon [19].

Here we describe how this was done for subsumption ($\sqsubseteq$), conjunction ($\sqcap$), negation ($\neg$), existential quantification ($\exists$), and universal quantification ($\forall$), i.e., most of the basic Description Logic language $\mathcal{ALC}$ [3], which is a proper fragment of the OWL 2 DL ontology language that is a relatively popular and standardised input for CNLs and NLG [5,17].

Runyankore verbalizations were derived by analyzing the similarities in factors. In both languages, the main variables that affect verbalization are the

Table 2. A few constructors, their typical verbalization in English, basic options in isiZulu and Runyankore; columns 1–3 were obtained from [14].

DL	English	Verbalization options isiZulu	Verbalization options Runyankore
⊑	... is a ...	Depends on what is on the rhs of ⊑; requires either a semantic (living vs non-living thing) or syntactic (noun starts with either i or a, o, u) distinction	Depends on whether the noun on the RHS of ⊑ starts with a vowel
≡	(1) ... is the same as ... (2) ... is equivalent to ...	I. Depends on what is on the rhs of ≡: whether a person or not II. Depends on grammatical number on lhs of ≡: whether singular or plural	I. Depends on what is on the LHS of ≡ to obtain the subject prefix II. Depends on whether what is on the RHS of ≡ starts with a vowel or not
⊔	... or ...	(1) ... okanye ... (2) ... noma ...	... nainga ...
⊓	... and ...	Depends on whether ⊓ is used to enumerate lists or connect clauses	Also depends on whether ⊓ is used to enumerate lists or connect clauses
¬	not ...	angi/akusiso/akusona/akubona /akulona/....	Depends on: I. both nouns, with the noun after ¬ dropping its initial vowel II. on the noun class of the concept in the relation
∃	(1) some ... (2) there exists ... (3) at least one ...	Depends on position in axiom: I. quantified over concept, depends on meaning of concept (living or non-living) II. includes relation (preposition issue omitted)	Depends on the noun class of the concept quantified over
∀	(1) for all ... (2) each ...	Depends on the semantic distinction of what is quantified over: I. (non)living thing II. noun class distinction	Depends on the noun class of the concept quantified over

noun class of a concept's name, the concept name's category, whether the concept is atomic or an expression, the quantifier, and the position of the concept in the axiom [14]. The isiZulu verbalization patterns were then changed to reflect Runyankore grammar rules. Table 2 depicts the similarities and customizations. The enumerations in the isiZulu and Runyankore columns indicate that the use depends on the context, which may be the category or noun class it applies to, or other aspects in the axiom before or after the symbol [14].

We illustrate the grammar rules associated with the verbalization of the selected constructors using examples of verbalizations. Although these grammar rules are different for isiZulu and Runyankore, it is important to keep in mind that the resulting Runyankore verbalizations were obtained from analyzing how it was done in isiZulu. For some constructors, there were several possible alternative verbalizations. The ones presented below were selected as the best during the survey (described in Sect. 5).

Universal Quantification. In isiZulu, the 'all' or 'each' uses the same translation, *-onke* [13,14]. Runyankore, however, has separate translations: *-ona* 'all' and *buri* 'each', with the latter only for nouns in the singular and the noun drops the initial vowel. In both isiZulu and Runyankore, the *-onke* and *-ona*, respectively, are prefixed with the appropriate prefixes of the noun class of the named concept (oral prefix for isiZulu [13] and genitive for Runyankore). However, in the isiZulu pattern, *-onke* is placed before the noun [13,14], while Runyankore places *-ona* after the noun. The example illustrates verbalising $Girl \sqsubseteq ...$ with 'each' (1z, 1r) and 'for all' (2z, 2r), including vowel processing (e.g., *-a+o-* = *-oo-* in Runyankore):

1z: isiZulu: <u>*Wonke*</u> *umfana* ...   (from *u-* + *-onke*)
1r: Runyankore: <u>*Buri*</u> *mwishiki* ...   (always *buri*)
2z: isiZulu: <u>*Bonke*</u> *abafana* ...   (from *ba-* + *-onke*)
2r: Runyankore: *Abishiki* <u>*boona*</u>   (from *ba-* + *-ona*)

Simple Taxonomic Subsumption. Verbalizations of $\sqsubseteq$ in Runyankore and isiZulu both depend on the first letter of the superclass. In isiZulu, the right copulative was selected based on the first letter of the noun of the superclass (*ng* for nouns starting with a-, o-, or u-, else *y*) [13]. In Runyankore, *ni* is used if the superclass starts with a consonant, and *n'* otherwise. For example, the verbalization of $Giraffe \sqsubseteq Animal$ ('each giraffe <u>is an</u> animal'):

   isiZulu: *indlulamithi <u>yi</u>silwane*
   Runyankore: *entwiga <u>n'</u>enyamishwa*

   If the subsumption is followed by negation, then the verbalization changes for both isiZulu and Runyankore. In isiZulu, the verbalization for subsumption and negation are combined into one term and the copulative is omitted, regardless the quantifiers in the verbalization [13]. Runyankore simply replaces *ni* for *ti* ('is not'); the noun after $\neg$ drops its initial vowel if it has one. This is illustrated for $Cup \sqsubseteq \neg Glass$ ('each cup <u>is not a</u> glass'):

   isiZulu: *zonke izindebe <u>aziyona</u> ingilazi*   (preferred verbalisation [14])
   Runyankore: *Ekikopo <u>ti</u> girasi*

Conjunction. isiZulu verbalizes $\sqcap$ depending on whether 'and' is used for a list of things or to connect clauses; in the former case, *na* is used and *kanye* or *futhi* for the latter [13]. Runyankore follows a similar pattern by using *na* when $\sqcap$ is used to enumerate lists, and *kandi* when used to connect clauses. Runyankore makes a further distinction for lists and uses *na* only when $\sqcap$ is between nouns,

Algorithm 4.1. Verbalization of Conjunction ($\sqcap$)

1: A axiom; Variables: a_1, a_2, p_1, p_2, c_1, sp, ap, a'; and Functions: $getLHS(A)$,
   $getRHS(A)$, $getPOS(a)$, $getNC(a)$, $getPreviousNoun(A)$
2: $a_1 \leftarrow getLHS(A)$ {get element to the left of $\sqcap$}
3: $a_2 \leftarrow getRHS(A)$ {get element to the right of $\sqcap$}
4: $p_2 \leftarrow getPOS(a_2)$ {get the part of speech for a_2}
5: **if** $p_2 = noun$ **then**
6: Result $\leftarrow$ 'a_1 na a_2'   {Verbalize with 'na', with vowel assimilation}
7: **else**
8: $c_1 \leftarrow getNC(a_1)$
9: $ap \leftarrow getAdjectivePrefix(a_1)$
10: Result $\leftarrow$ 'a_1 kandi apa_2'   {Verbalize with 'kandi'}
11: **end if**
12: **return** Result

---

but *kandi* otherwise. If one of the concepts is an adjective, then the noun class is required to obtain the adjective prefix in order to form the full translation of the adjective. Algorithm 4.1. illustrates the verbalization of $\sqcap$, where the first concept is a noun and the second is either a noun or adjective.

Existential Quantification. In both isiZulu and Runyankore, the noun class is crucial to the verbalization: in obtaining the relative and quantitative concords in isiZulu, and the subject prefix in Runyankore. Runyankore verbalizes $\exists$ as *hakiri* for 'at least' and *-mwe* with subject prefix of the concept quantified over

Algorithm 4.2. Verbalization of Existential Quantification ($\exists$)

1: A axiom; Variables: a_1, a_2, sp_1, r, r'_1, r'_2, sp_r, c_r, c_1; and Functions: $getConcept(A)$,
   $getRole(A)$, $getRoleElement(r)$, $getNC(a)$, $dropInitialVowel(a)$, $splitRole(r)$
2: $a_1 \leftarrow getConcept(A)$ {get the concept quantified over in the axiom}
3: $r \leftarrow getRole(A)$
4: $a_2 \leftarrow getRoleElement(r)$ {get the element of the role}
5: $c_1 \leftarrow getNC(a_1)$
6: $sp_1 \leftarrow getSubjectPrefix(c_1)$
7: $a'_1 \leftarrow dropInitialVowel(a_1)$
8: **if** $r.hasForm(hasX) = true$ **then**
9: $splitRole(r)$ {split the role into its constituent terms}
10: $r'_1 \leftarrow r[0]$ {the role}
11: $r'_2 \leftarrow r[1]$ {the named concept in the role}
12: $c_r \leftarrow getNC(r'_2)$
13: $sp_r \leftarrow getSubjectPrefix(c_r)$
14: Result $\leftarrow$ 'Buri a'_1 hakiri $sp_1 r'_1$e r'_2 sp_rmwe sp_rri c_2'
15: **else**
16: Result $\leftarrow$ 'Buri a'_1 nisp_1ra hakiri a_2 sp_2mwe' {Verbalize with 'hakiri ... -*mwe*'}
17: **end if**
18: **return** Result

Algorithm 4.3. Verbalization of Negation ($\neg$) for 'hasX'

1: A axiom; Variables: a_1, a_2, r, r_1, r_2, c_1, c_r, sp_1, sp_r; and Functions: $getRole(A)$, $getConcept(A)$, $getNC(a)$
2:   $a_1 \leftarrow getConcept(A)$ {get the concept at the start of the axiom}
3:   $r \leftarrow getRole(A)$ {get the role after $\neg$}
4:   $c \leftarrow getNC(a_1)$
5:   $sp_1 \leftarrow getSubjectPrefix(c)$
6: **if** $r.hasForm(hasX) = true$ **then**
7:     $splitRole(r)$ {split the role into its constituent terms}
8:     $r_1' \leftarrow r[0]$ {the role}
9:     $r_2' \leftarrow r[1]$ {the named concept in the role}
10:     $c_r \leftarrow getNC(r_2')$
11:     $sp_r \leftarrow getSubjectPrefix(c_r)$
12:     Result $\leftarrow$ ' ... ti$sp_1 r_1'$e r_2' sp_rri ... '   {Verbalize with 'ti' to negate 'has' and the subject prefix of the named concept in the role}
13: **else**
14:     Result $\leftarrow$ ' ... tisprikua_2a ... '   {Verbalize with 'ti' which negates verbs}
15: **end if**
16: **return** Result

in order to form the full word for 'one'. The latter is similar to isiZulu's -*dwa*, which also relies on the noun class to get the correct concords [13].

> isiZulu: *wonke uSolwazi ufundisa isifundo* <u>*esisodwa*</u>   (but pl. preferred [14])
> Runyankore: *Buri mwegyesa nayegyesa* <u>*hakiri*</u> *eishomo* <u>*rimwe*</u>

Algorithm 4.2. includes verbalization of the verb in 3rd pers. sg. and 'hasX' named roles (e.g., *hasChild*).

Negation with Roles. Negation of verbs also uses *ti* (as is the case for subsumption), but additionally requires the subject prefix of the concept and the infinitive *ku*. In the case of 'hasX'-named roles, the subject prefix of X—the concept contained within the role name—is used. Algorithm 4.3. shows how this has been achieved for both cases.

5   Evaluation

The verbalizations of $\forall$, $\sqsubseteq$, $\neg$, and $\exists$ produced several alternative texts during the initial analysis, as they did for isiZulu. In order to decide which verbalization to implement, an evaluation similar to that of [9,14] was carried out to ascertain which pattern was preferred by survey participants.

Survey Design. The participants in the evaluation were obtained from Runyankore speakers in Kampala, Uganda. WhatsApp was used to conduct the survey because it is more familiar and widely used than email or online surveys.

18 participants completed the survey, who were middle-class Banyankore and spoke both English and Runyankore; 78.8 % were female and their age ranged from 24 to 59. Participants were recruited using snowball sampling, starting with a family WhatsApp group. They were instructed to answer five questions by subjectively selecting the best verbalization for each question in the form 1c, 2d, 3a, etc. They were also encouraged to explain the reasons for their choices, though only 3 participants did so. The answers were either directly delivered to us via WhatsApp, or were sent through an intermediary. Nineteen participants replied, though one was excluded because it was incomplete.

The five questions in the survey tested verbalisations of the following axioms: (1) $Teacher \sqsubseteq \exists teaches.Subject$; (2) $Cup \sqsubseteq \neg Thing$; (3) $Cat \sqsubseteq Animal$; (4) $Man \sqsubseteq \exists hasChild.Doctor$; and (5) $Giraffe \sqsubseteq \neg\exists eats.Meat$. The English verbalization was included before the Runyankore alternatives in order to enable the participants to translate to Runyankore, and then compare with the alternatives generated from the verbalization patterns.

Question (1) had four alternatives, with the difference being either the singular or plural form, as well as the placement of *hakiri* before or after the noun. Question (2) had four alternatives: singular with *ti*, plural with *ti*, singular with *ti... ri*, and plural with *ti... ri*. Question (3) had two alternatives: either singular or plural. The placement of *hakiri* before or after the verb, the choice of whether to include *-mwe*, and either singular or plural resulted in six alternatives for question (4). Question (5) also had six alternatives, due to singular/plural, and the difficulty of translating $\neg\exists eats$ as 'never eats', 'does not eat', or 'is not eating'.

Results. The singular form was generally preferred by the majority of survey participants. 72.2 % chose the singular form for $\exists$ with *hakiri* after the verb; 55.6 % preferred the singular with *ti* for $\sqsubseteq \neg$, and the singular with $\sqsubseteq$ was preferred by 72.2 % of the participants. The plural form was only preferred in the case of negating a verb ($\neg\exists eats$) by 33.3 %. Three verbalizations were not selected by any participants: question 4, the plural form with *hakiri* after the verb, as well as the plural with no *-mwe*' and question 5, the singular form with the translation as 'does not eat.' As no explanations were offered by the participants, we cannot speculate as to the reasons for this.

The results were not as clear for individual choices for $\exists hasChild.Doctor$ as the singular with *hakiri* before or after the verb, and the plural with *hakiri* before the verb were all chosen by 27.8 % of participants. However, when generalized along singular/plural lines, then 72.2 % selected the singular; when based on the placement of *hakiri* and the inclusion of *-mwe*, then the most preferred by 44.4 % was the one which had *hakiri* before the verb and included *-mwe*.

The evaluation of $\neg\exists eats$ also produced interesting results. The plural form for the translation as 'never eat' was chosen by the majority of participants (33.3 %). However, because 'never' verbalizes the axiom as if it has a temporal dimension, which OWL does not have, we considered the second best alternative, which was the plural form of the translation as 'are not eating' (27.8 %). Despite this, the evaluation still showed that the plural is preferred for negation of roles.

6   Discussion

Kuhn [15] stated that one of the applications of CNLs is to provide natural and intuitive representations of formal languages. Verbalization is one of the ways of doing this, and our work here further solidifies what was done in [13,14] to show that it is possible for Bantu languages, despite their complex linguistic structure. From our evaluation, we are able to control for certain factors during verbalization, such as the grammatical form as well as the placement of *hakiri* and the inclusion of *-mwe* in the resulting text, thus reducing the number of construction and interpretation rules. This is important because it ensures a deterministic outcome of the resulting CNL. Additionally, the subset of Runyankore applied during verbalization is restricted by the types of axioms that can be represented in $\mathcal{ALC}$. This further ensures that a predictable interpretation is obtained, and this can only be done by having a strict syntactic subset of natural language [8,9].

The application of the bootstrapping approach in the development of language resources, especially for similar languages, has been mainly associated with reduction in development time and effort. This paper further highlights its importance to under-resourced languages like Runyankore. Tailoring Runyankore verbalizations from those of isiZulu made it a lot easier, because the underlying theory, like the factors affecting verbalization and the role of the noun class, had already been identified for isiZulu. Table 2 showed that verbalization in both languages is affected by similar factors, and Sect. 4 showed how the tailoring was done to obtain Runyankore verbalizations. This jumpstarting also facilitated extending the verbalisations with 'hasX' named roles. Our Java implementation of the above algorithms also creates the possibility to verbalize longer axioms.

Further, with the development of an ontology based on the Bantu noun class system—for annotating an ontology with noun class information [6]—there is now the possibility of applying such verbalizations to real-world use cases. Our research shows that the customization of these patterns is not only possible, but can obtain good results, as is the case in our evaluation. The same approach can thus be applied to languages like Rukiga, Rutooro, and Runyoro that are regarded to be between 78 % and 99 % similar to Runyankore [2,18,20]. Finally, given that a bootstrapping approach was still possible between Runyankore and isiZulu, which are in different zones according to Guthrie's classification of Bantu languages [10,16], the same approach could be feasible even for those Bantu languages which are not classified under the same category.

7   Conclusions

The verbalization of most $\mathcal{ALC}$ constructors in Runyankore was made easier by applying the bootstrapping approach to what was done for isiZulu. We have identified that verbalization in Runyankore is affected by the same factors as in isiZulu, namely: the noun class of the name of the concept, the category of the concept, whether the concept is atomic or an expression, the quantifier used in the axiom, and the position of the concept in the axiom. A few differences were identified, so the Runyankore ones were tailored to this. The evaluation by non-linguists provided clear favorites among alternative verbalization options. We plan to complete an evaluation among linguists concerning grammatical correctness, as well as a more inclusive sample of participants along age, education, and socio-economic lines.

Acknowledgements. This work is based on the research supported by the Hasso Plattner Institute (HPI) Research School in CS4A at UCT and the National Research Foundation of South Africa (Grant Number 93397).

References

1. Angelov, K., Ranta, A.: Implementing controlled languages in GF. In: Fuchs, N.E. (ed.) CNL 2009. LNCS, vol. 5972, pp. 82–101. Springer, Heidelberg (2010)
2. Asiimwe, A.: Definiteness and specificity in Runyankore-Rukiga. Ph.D. thesis, Stallenbosch University, Cape Town, South Africa (2014)
3. Baader, F., Calvanese, D., McGuinness, D.L., Nardi, D., Patel-Schneider, P.F. (eds.): The Description Logics Handbook - Theory and Applications, 2nd edn. Cambridge University Press, Cambridge (2008)
4. Bosch, S., Pretorius, L., Fleisch, A.: Experimental bootstrapping of morphological analyzers for Nguni languages. Nordic J. Afr. Stud. **17**(2), 66–88 (2008)
5. Bouayad-Agha, N., Casamayor, G., Wanner, L.: Natural language generation in the context of the semantic web. Semant. Web J. **5**(6), 493–513 (2014)
6. Chavula, C., Keet, C.M.: Is lemon sufficient for building multilingual ontologies for Bantu languages? In: Proceedings of OWLED 2014, CEUR-WS, vol. 1265, pp. 61–72, Riva del Garda, Italy (2014)
7. Davel, M., Barnard, E.: Bootstrapping in language resource generation. In: Proceedings of PRASA 2003, Langebaan, South Africa (2003)
8. Gruzitis, N., Barzdins, G.: Towards a more natural multilingual controlled language interface to owl. In: 9th International Conference on Computational Semantics (IWCS), pp. 335–339 (2011)
9. Gruzitis, N., Nespore, G., Saulite, B.: Verbalizing ontologies in controlled baltic languages. In: Proceedings of International Conference on HLT-The Baltic Perspective, FAIA, vol. 219, pp. 187–194. IOS Press, Riga, Latvia (2010)
10. Guthrie, M.: The Classification of the Bantu Languages. Oxford University Press, London (1948)
11. Jarrar, M., Keet, C.M., Dongilli, P.: Multilingual verbalization of ORM conceptual models and axiomatized ontologies. Vrije Universiteit, Brussels, Belgium, Technical report (2006)
12. Joubert, L., Zimu, V., Davel, M., Barnard, E.: A framework for bootstrapping morphological decomposition. In: Proceedings of PRASA 2004, Grabouw, South Africa (2004)
13. Keet, C.M., Khumalo, L.: Basics for a grammar engine to verbalize logical theories in isiZulu. In: Bikakis, A., Fodor, P., Roman, D. (eds.) RuleML 2014. LNCS, vol. 8620, pp. 216–225. Springer, Heidelberg (2014)

14. Keet, C.M., Khumalo, L.: Toward verbalizing ontologies in isiZulu. In: Davis, B., Kaljurand, K., Kuhn, T. (eds.) CNL 2014. LNCS, vol. 8625, pp. 78–89. Springer, Heidelberg (2014)

15. Kuhn, T.: A survey and classification of controlled natural languages. Comput. Linguist. **40**(1), 121–170 (2014)

16. Maho, J.F.: Nugl online: The online version of the updated guthrie list, a referential classification of the bantu languages (2009). http://goto.glocalnet.net/mahopapers/nuglonline.pdf

17. Safwat, H., Davis, B.: CNLs for the semantic web: a state of the art. Lang. Resour. Eval. 1–30 (2016, in print). doi:10.1007/s10579-016-9351-x

18. Tayebwa, D.D.: Demonstrative determiners in Runyankore-Rukiga. Master's thesis, Norwegian University of Science and Technology, Norway (2014)

19. Taylor, C.: A Simplified Runyankore-Rukiga-English Dictionary. Fountain Publishers, Kampala (2009)

20. Turamyomwe, J.: Tense and aspect in runyankore-rukiga: linguistic resources and analysis. Master's thesis, Norwegian University of Science and Technology, Norway (2011)

Statistically-Guided Controlled Language Authoring

Susana Palmaz[(✉)], Montse Cuadros, and Thierry Etchegoyhen

Vicomtech-IK4, Donostia - San Sebastián, Spain
{spalmaz,mcuadros,tetchegoyhen}@vicomtech.org
http://www.vicomtech.org

Abstract. This study presents a series of experiments using contextual next word prediction to aid controlled language authoring. The goal is to assess the capabilities of n-gram language modelling to improve the generation of controlled language in restricted domains with minimal supervision. We evaluate how different dimensions of language model design can impact prediction and textual coherence. In particular, evaluations of suggestion ranking, perplexity gain and language model combination are presented. We show that word prediction can provide adequate suggestions which could offer an alternative to costly manual configuration of rules in controlled language applications.

Keywords: Word prediction · Controlled language · Language modelling

1 Introduction

The efficient creation of coherent texts in professional technical writing environments presents a series of challenges that can be tackled with the application of controlled language techniques. These methods define an accepted set of heuristics that are incorporated into the authoring process through control systems which aid the technical writer in the process of editing texts that reflect the selected linguistic traits of the domain. The required degree of control will depend mainly on the goals that the final text needs to meet and on the scope of the controlled language rules. Controlled language specifications can span from highly-controlled formulaic syntactic frames to more informal choices that represent expressive preferences at the discourse level. Despite the heuristics approach being the most common method for controlled language generation, completely automatic alternatives have not yet been explored in full.

In this work a series of experiments are presented to evaluate the power of statistical contextual predictive typing to aid in the process of creating linguistically coherent texts. The final goal is to assess what gains can be obtained in this setting towards the creation of easily trainable controlled language environments. Word prediction has shown good results in keystroke reduction for assistive typing systems [1,2] where the main goal is to help the user reduce the physical effort derived from typing. The assumption here is that technical writing

© Springer International Publishing Switzerland 2016
B. Davis et al. (Eds.): CNL 2016, LNAI 9767, pp. 37–47, 2016.
DOI: 10.1007/978-3-319-41498-0_4

can potentially benefit from the increased typing performance features offered by the generation of in-domain contextual suggestions and the provision of an indirect unsupervised controlled language system. The influence of traditional language modelling features over the generation of appropriate suggestions will be measured, in relation to suggestion ranking and text coherence.

The presentation is structured as follows: a review of current relevant approaches will be presented first, followed by the design of the system and the experimental approach. The experimental results will then be analysed. Finally, the conclusions of this work will be summed up with a proposal for future work.

2   Background

Controlled natural languages are subsets of natural languages that display reduced or minimal ambiguity and complexity. The specification of the language is made through explicit restrictions on the grammar, lexicon and style that limit the applicability of general rules of the language. This broad definition has made it difficult to find general agreement on the characteristics of these languages and has generated different methodological approaches, linked to the particular application intended to exploit them. In general terms, a major distinction can be drawn from whether the language is intended to be consumed or produced by humans or machines, but additional features can be considered, e.g., whether the language represents a general or restricted domain, or whether readability, parsability or translatability is the focus of the final text. An in-depth survey and classification of controlled language types is offered in [3].

A recurrent application of these technologies has been within organizations that wish to unify their linguistic output. Particularly in the technical domain, having means to help professional technical writers produce texts that are linguistically consistent and unambiguous can have several benefits derived from improvements in readability and comprehension, which can be critical if the documents are to be shared within or outside the company. Controlled Languages in technical documentation have been applied in a wide variety of industrial documentation applications, such as aircraft design at Airbus [4] or language standardisation at Ford Motor Company [5]. Controlled languages also provide a strong foundation for the translation of corporate documentation [6].

However, the advantages of using controlled languages come at the cost of manually developing the knowledge specific rules, restricted vocabularies and a knowledge representation for each specific corporation or domain of application.

Word prediction is a collection of methods designed to generate text completion suggestions given some input. This technology was proposed in the context of speech recognition to improve the selection of recognition hypotheses [7]. It was later adopted and further developed in the integration of typing assistants, either destined to typing aids for people with disabilities or, more recently, to predictive typing features for mobile phones and information retrieval applications [8]. The nature of the generated suggestions (words, words plus usage examples, grammatical categories, etc.) and the strategy used to exploit the available contextual information (length of context, depth of linguistic analysis, etc.) determine the design of the system and the potential gains that can be extracted from it [8].

The most simplistic word prediction systems available use word frequency and the character context already typed by the user to predict the ending of the current word. Despite the simplicity of this method and its ability to ease input effort to a certain extent, the predictive element of this approach is limited and there is no guarantee that the offered suggestions will match the context linguistically. Many lightweight writing aid tools in the market are limited to this functionality, like Sumitsoft's Typing Assistant[1] or code completion features from integrated development environments, to give two examples.

Contextual word prediction goes one step further to improve the generation of suggestions through the analysis of different elements of the context. N-gram language models based on words or words plus part-of-speech are the most commonly used approaches in this setting, exemplified by systems like WordQ[2] and projects like FASTY [9], although other modelling alternatives have been proposed like LSA [10], decision trees [11], and non-linear collocation features [12]. An in-depth revision of text prediction systems is offered in [8].

From the point of view of prediction quality, studies have shown how performance is dependent on traditional language modelling factors such as n-gram size, training data size, data sparseness handling and similarity between training and target domains [2,11,13]. Additionally, the goal of the system as a real-time suggestion generator requires it to perform queries and calculations in a limited period of time and to avoid typing degradation due to the cognitive load derived from scanning too many or wrong suggestions [8,11].

In general, from the perspective of word prediction applied as a method to improve typing efficiency, evaluations have shown that increases in communication rates can be obtained with this method. This work will evaluate whether these improvements also extend towards increased coherence in the produced texts when predictions are mediated by a language model from a particular domain. Apart from the potential of this method to offer a cheap alternative to manual maintenance of vocabularies and rule sets for controlled languages, the ability of this method to be objectively evaluated offers valuable insights into the predictive power of traditional language modelling techniques.

Although controlled language applications and next word prediction have developed into mature implementations in recent years, few efforts have been made in the direction of merging these two technologies. Some controlled language implementations have attempted to integrate look-ahead functionality to improve user performance [14–16], aiding the user by generating grammatically correct suggestions based on a particular grammar. These systems present an interesting approach to controlled language generation, but prediction performance has not been formally evaluated yet and cannot be easily compared with our current methodology and results.

---

[1] http://www.sumitsoft.com/.
[2] http://www.goqsoftware.com/wordQ.php.

3   Approach

The work presented here has been created with an n-gram word prediction module developed by Vicomtech-IK4. This module is part of a library that was created as an experimental, easily-trainable, controlled language environment generator tool for a company that offers technical writing services for industrial businesses. Their request was to create a system that could be automatically trained from their existing data and which could offer means to improve the efficiency of their technical writers at maintaining domain-specific textual coherence with minimal oversight.

The complete system developed for this task includes a full textual extraction and normalization pipeline to generate the domain corpora for the system, automatic terminology extraction, n-gram language modelling, and word prediction as its main components. This system is currently being evaluated by technical writers to assess the implemented functionality, the results of which will be evaluated as they become available. The objective evaluation shown here has been performed as a formal in-house assessment of the word prediction module.

Regarding the design features of this module, a few elements were considered:

- Speed: predictions need to be generated fast to avoid lag in user performance.
- Context processing:
  - Directionality: left-only or left and right context.
  - Character context: the currently typed prefix of a word can be added to the calculation to filter results.
  - Hypothesis generation strategy: exhaustive search of all known contexts or longest productive context.
- Terminology: identify domain-specific terminology in suggestions
- Multiple model combination: linear combination of language models for multifaceted domain characterization.

4   Experiments

An objective observation of the effects of training and context size on rank distribution and information gain was performed. The experimental approach was inspired by the results presented in [11]. Here the notion of suggestion rank evolution is introduced to expand the analysis of token prediction accuracy and evaluate the effects of different language modelling features on prediction beyond top suggestion results.

We focus on analyzing rank as a defining factor of the expected quality of the system instead of using more traditional measures of keystroke reduction rates to get an objective observation of the predictive power of the system from the point of view of the user. Different results in assistive word prediction have shown how the quality of the suggestions is critical for the system to achieve effort reduction [2]. With a similar intuition, we expect good ranking to play an essential role in indirectly conditioning the users into writing according to the style of the domain as determined by the language model.

4.1   Experimental Setup

The data used in the experiments consists of two individual corpora, one in-domain and the other generic. The domain corpus was chosen to be the English portion of the European Medical Agency corpus (EMEA) [17], downloaded from the OPUS project website[3]. This corpus is a fair representative of a restricted domain, in this case medical, and can be accessed openly while offering a decent size for language modelling experiments. From an initial data drop of approximately one million sentences, nearly 285000 sentences were considered after excluding duplicates. The generic corpus was a manually created combination of news, excerpts from books from Project Gutenberg and Europarl for a total of 250000 sentences with similar data normalisation criteria.

All corpora were tokenised and truecased using scripts from Moses [18]. From the resulting domain corpus 2850 sentences (1 %) were extracted randomly for testing and the remaining data was used for training. This amount of test sentences was chosen to keep as much data as possible for training models while keeping a usable test size. Additionally, sentences shorter than the maximum n-gram size were discarded, with a final training size of 238332 sentences. All language models were estimated with the KenLM library [19] using modified Kneser-Ney smoothing without pruning.

The context processing method in all cases was similar. Only left contexts of sizes up to four were considered. The hypothesis generation strategy implemented did not consider all contexts to improve efficiency, and did so by iteratively decreasing the context around the prediction target until a span returns results. This method was introduced to bias prediction in favour of suggestions generated with the longest possible context. This allowed the algorithm to work efficiently against real time performance with a middle sized 5-gram model.

The evaluation of rank was based on predicting suggestions for all tokens in the test corpus sentence by sentence using a sliding window of a defined context size and locating the correct word in the list of suggestions. Then the ranks of the validated suggestions are aggregated in bins to track the distribution of the correct target words in a spectral representation. Figures 1, 2 and 3 display this analysis where rank bins are identified by: (1) top rank, (2) ranks 2nd to 5th (3) ranks 6th to 10th (4) the end of the list ("rest" in the legend) (5) predicted suggestions that did not include the correct target ("NF") (6) contexts that did not generate any suggestions ("NS"). The grey-scale gradient was chosen to represent rank suitability for word prediction, where top ranks are preferred.

4.2   Results

Context Size. Figure 1 shows the effect of left context size on suggestion ranking when predicting from the in-domain language model for context sizes one (bigram prediction) to four (5-gram). The graph clearly shows how performance improves for each additional level and how the improvements are not linear.

---

[3] http://opus.lingfil.uu.se/.

The traditional trigram approach that was predominant in word prediction can also be seen here in comparison with the performance gains of considering orders of up to 5-grams.

It is also important to observe here that although the hypothesis generation strategy is not search-exhaustive to favour speed (only the longest productive context is considered in every prediction), it shows the expected improvements related to context. In the case of context size four (Fig. 2b) the system predicted the correct target in the top rank 51.1 % of the cases and within the top 5 positions 64.6 %. In contrast, as less context is made available to the system the rank classification power becomes increasingly diluted. At the same time, the hypothesis generation strategy decreases the chances of finding the correct next word when a long context is productive and more predictions lacking the target word are generated.

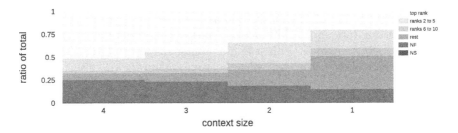

Fig. 1. Rank distribution of correct targets with a model trained on 238332 sentences.

Training Size. The graphs in Fig. 2 complement the analysis of context size by showing the evolution of rank when different sizes of training data are used. Results for context sizes of one (Fig. 2a) and four (Fig. 2b) are presented. The comparison of the two graphs clearly shows that adding more training data does not improve the performance for both in the same way beyond a certain size. While 5-gram prediction improves with additional data consistently, bigram prediction does not improve at the same rate as it lacks the classification power to rank results appropriately. Except for improvements in vocabulary coverage, which can be tracked by the trend to reduce the number of not-classified targets, rank distribution of bigram prediction stagnates quickly. We can also see the interaction of the hypothesis generation strategy and context size. This effect is also reduced when more training data is used. A more thorough study of the type and impact of these failed predictions would be necessary to optimize the implementation.

Character Context Size. Apart from using the left word context for predictions, additional experiments were made to use the current character context prefix to filter suggestions further. Results are presented in Fig. 3, where we compare the performance of contexts of size one (Fig. 3a) and four (Fig. 3b).

Adding character context is positive for bigram prediction and performance is boosted from a mere 20.4 % of top ranked correct suggestions (40.4 % within the

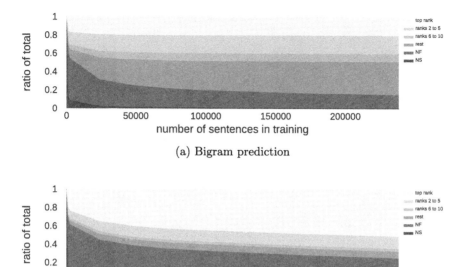

(a) Bigram prediction

(b) 5-gram prediction

Fig. 2. Evolution of rank distribution with different training sizes

top five) to 40.9 % with a character context of two (53.4 % within the top five). In contrast, prediction with 5-grams shows similar but less dramatic improvements when using the first character (top result raises from 51.1 % to 58.2 % for the top result and 64.6 % to 65.7 % for the top five), but performance is degraded when using two characters (51.2 % top result, 56.3 % top-five), to the point of showing a surprisingly close distribution to bigram prediction of the top five results with the same character context available (bigram 53.9 %, 5-gram 56.3 %).

This difference is attributed to the filtering capacity imbalance between word and character context. Bigrams are benefited in this setting as they are less restrictive in the hypotheses they generate which are subsequently filtered with the character prefix. Although this result is interesting to show modelling limitations and potential implementation nuances, the expectation on the user to know the first two letters of the target word might not be ideal in a controlled language environment in all cases. A user-based evaluation will be able to offer a more practical insight into the effects of character context.

Perplexity Analysis. Next we discuss results for model perplexity obtained from texts corrected automatically using word prediction. The methodology was similar to the approach followed above, but instead of analysing rank, the target word is substituted by the top suggestion and printed to an output file. When all predictions are generated, the distance between the perplexities of the original and predicted texts is calculated. The intuition is that as the original test

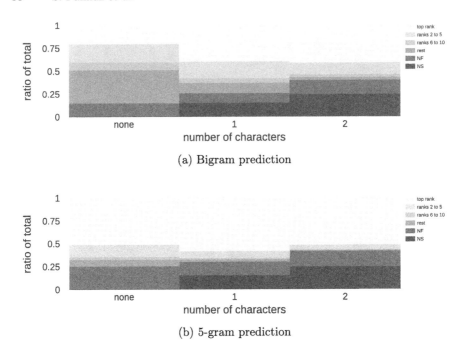

(a) Bigram prediction

(b) 5-gram prediction

Fig. 3. Effects of adding character context to word context prediction

data is a correct controlled language target and an ideal word prediction system would be one which produces the same text from the top results with a resulting perplexity distance of zero. To avoid dragging the increment in lexical distance between the documents during the process, the generated substitutions are only replaced in the predicted text, while prediction contexts are taken from the original data. This is an objective measure which is less optimal than usability tests, but can nonetheless give an automatic mean of testing results. A more thorough test design could expand these results to observe how different contexts are resolved with this system.

The graph in Fig. 4 shows the evolution of perplexity distance as training data is modelled at different sizes for contexts of one to four words. The results here match the expectation demonstrated by rank results. Bigram prediction is not able to offer good enough ranking without additional effort from the user to scan the list and find the right suggestion away from the top position, and this effect only worsens with larger models. As longer context sizes are considered, the amount of noise is reduced to the point where 5-gram prediction is able to progressively close the perplexity gap with the original document as more data is added to the model.

Model Combination. Finally, preliminary results for linear model combination are presented. Figure 5 shows the effect of adding three in-domain models of different sizes to a generic model baseline and observing how weighting influences

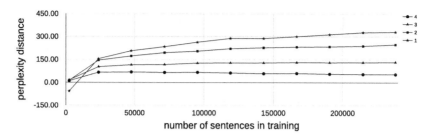

Fig. 4. Training size impact on perplexity reduction for contexts of one to four words

token prediction accuracy. By changing the size of the models in this test we are reproducing usage scenarios where little in-domain data is available.

Using generic models to improve the performance of small restricted domain's models is intuitive and common practice. The theoretical justification for this is that a generic model will be able to capture contexts present across language manifestations within a language regardless of domain restrictions. This will in turn improve coverage and give support to the domain model. The graph shows that this might not necessarily be justified for word prediction in all cases.

Three training data sizes were chosen: a (forcibly) small model trained on 1192 sentences (0.5 % of the available training data), a more realistic but still small model built from 23833 sentences (10 %) and the model trained on the maximum number of available training sentences (238332). Looking at the baseline performances of the three domain models for prediction with 5-grams, even the exceedingly small one, all of them offer token prediction accuracies far superior to those of the generic model when predicting on their own (8.3 % for the baseline model versus 20.2 %, 33.5 % and 47.7 %). Only for the smallest model can a slight improvement be observed when the weight is adjusted to around 0.8 (from a 20.2 % baseline to 20.6 %). Such a small amount of training data is unrealistic for a real world word prediction system, but the result is interesting nonetheless as it shows how domain data (regardless of size) is able to successfully model generic contexts as well as the specific domain's idiosyncrasies without additional aid.

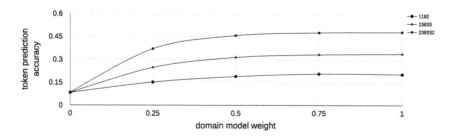

Fig. 5. In-domain model weighting for different domain training sizes

A more thorough observation of the effects of model combination with different generic models would be necessary to evaluate the potential improvements that model combination can yield with this method.

5   Conclusions and Future Work

We presented an objective evaluation on the use of contextual word prediction, determining its potential to enhance controlled language authoring. The ability of language models to produce appropriately ranked suggestions according to domain statistics makes this approach an interesting source for easily-trainable, indirectly-guided controlled language environments. The results show the impact of language and context modelling features on prediction quality and the potential of using in-domain contextual predictions to generate productive suggestions for the creation of acceptable controlled texts. We expect to complement these objective results with user evaluations as they become available.

The work presented here is an ongoing effort. Additional features are being evaluated at the moment to improve several elements of the library, like multiword processing and additional classification features. From the point of view of evaluation, a similar analysis in other domains and other languages is intended, with particular attention to studying improvements through data selection and multiple model combination. Finally, we will complete our analysis with measures of the impact of domain-specific terminology, a key feature of the overall system we developed for controlled language authoring. Also, more advanced language modelling techniques [20] can be explored to improve prediction performance.

References

1. Koester, H.H., Levine, S.: Effect of a word prediction feature on user performance. Augmentative Altern. Commun. **12**(3), 155–168 (1996)
2. Trnka, K., McCaw, J., Yarrington, D., McCoy, K.F., Pennington, C.: User interaction with word prediction: the effects of prediction quality. ACM Trans. Accessible Comput. (TACCESS) **1**(3), 17 (2009)
3. Kuhn, T.: A survey and classification of controlled natural languages. Comput. Linguist. **40**(1), 121–170 (2014)
4. Spaggiari, L., Beaujard, F., Cannesson, E.: A controlled language at Airbus. In: Proceedings of EAMT-CLAW 2003, pp. 151–159 (2003)
5. Rychtyckyj, N.: Standard Language at Ford Motor Company: A Case Study in Controlled Language Development and Deployment. Cambridge University Press, Cambridge (2006)
6. Ramírez-Polo, L.: Use and evaluation of controlled languages in industrial environments and feasibility study for the implementation of machine translation. Ph.D. thesis, Universidad de Valencia (2012)
7. Wu, D., Sui, Z., Zhao, J.: An information-based method for selecting feature types for word prediction. In: EUROSPEECH (1999)
8. Garay-Vitoria, N., Abascal, J.: Text prediction systems: a survey. Univ. Access Inf. Soc. **4**(3), 188–203 (2006)

9. Trost, H., Matiasek, J., Baroni, M.: The language component of the FASTY text prediction system. Appl. Artif. Intell. **19**(8), 743–781 (2005)
10. Wandmacher, T., Antoine, J.Y.: Methods to integrate a language model with semantic information for a word prediction component. arXiv preprint (2008). arXiv:0801.4716
11. Van Den Bosch, A.: Scalable classification-based word prediction and confusible correction. Traitement Automatique des Langues **46**(2), 39–63 (2006)
12. Even-Zohar, Y., Roth, D.: A classification approach to word prediction. In: Proceedings of the 1st North American Chapter of the Association for Computational Linguistics Conference, NAACL 2000, Stroudsburg, PA, USA, pp. 124–131. Association for Computational Linguistics (2000)
13. Lesher, G.W., Moulton, B.J., Higginbotham, D.J., et al.: Effects of ngram order and training text size on word prediction. In: Proceedings of RESNA 1999, pp. 52–54 (1999)
14. Schwitter, R., Ljungberg, A., Hood, D.: Ecole-a look-ahead editor for a controlled language. In: EAMT-CLAW 2003, pp. 141–150 (2003)
15. Angelov, K., Ranta, A.: Implementing controlled languages in GF. In: Fuchs, N.E. (ed.) CNL 2009. LNCS, vol. 5972, pp. 82–101. Springer, Heidelberg (2010)
16. Kuhn, T.: A principled approach to grammars for controlled natural languages and predictive editors. J. Log. Lang. Inf. **22**(1), 33–70 (2013)
17. Tiedemann, J.: News from OPUS - a collection of multilingual parallel corpora with tools and interfaces. In: Nicolov, N., Bontcheva, K., Angelova, G., Mitkov, R. (eds.) Recent Advances in Natural Language Processing, vol. V, pp. 237–248. John Benjamins, Amsterdam/Philadelphia (2009)
18. Koehn, P., Hoang, H., Birch, A., Callison-Burch, C., Federico, M., Bertoldi, N., Cowan, B., Shen, W., Moran, C., Zens, R., et al.: Moses: open source toolkit for statistical machine translation. In: Proceedings of the 45th Annual Meeting of the ACL on Interactive Poster and Demonstration Sessions, pp. 177–180. Association for Computational Linguistics (2007)
19. Heafield, K.: KenLM: faster and smaller language model queries. In: Proceedings of the Sixth Workshop on Statistical Machine Translation, Edinburgh, Scotland, United Kingdom, pp. 187–197 (2011)
20. Jozefowicz, R., Vinyals, O., Schuster, M., Shazeer, N., Wu, Y.: Exploring the limits of language modeling. arXiv preprint (2016). arXiv:1602.02410

A Speech Interface to the PENG^{ASP} System

Christopher Nalbandian and Rolf Schwitter$^{(\boxtimes)}$

Department of Computing, Macquarie University, Sydney, NSW 2109, Australia
Christopher.Nalbandian@students.mq.edu.au, Rolf.Schwitter@mq.edu.au

Abstract. The increased presence and accessibility of online speech recognition services has encouraged an investigation of the effectiveness of using such a service to allow users to speak a textual specification in controlled natural language instead of typing it. Google's Web Speech API provides an accessible and portable speech recognition service that integrates well with web-based interfaces. Using Google's Web Speech API, we present the design and implementation of a speech-based interface for the PENG^{ASP} system. We do this in order to examine the usefulness of speech-based input for controlled natural language processing and to explore potential synergies between speech-based and text-based input.

Keywords: Controlled natural languages · Predictive text editors · Speech interfaces · Speech recognition services

1 Introduction

In recent years, the advances in speech recognition technology have resulted in the development of sophisticated online speech recognition services such as AT&T's Speech API[1], Google's Web Speech API[2], and IBM Watson's Speech to Text API[3]. The availability of these services enables the integration of speech-based input with user interfaces. In this context, it is interesting to investigate how useful speech-based input is for an existing controlled natural language processing system and to assess the effectiveness of speaking a controlled language instead of typing it. The controlled natural language chosen for this speech interface is the one of the PENG^{ASP} system [5] due to its base language being English, its expressiveness and its existing predictive text editor interface.

Controlled natural languages are defined as languages which have the properties of being based on precisely one natural language, being more restrictive than their base language through lexicon, syntax or semantics, preserving most of the properties of their base language and are constructed in such a way that the elements of the language are explicitly specified in a grammar [9,14]. An important feature of machine-oriented controlled natural languages is the ability to

---

[1] http://developer.att.com/apis/speech.
[2] https://www.google.com/intl/en/chrome/demos/speech.html.
[3] https://speech-to-text-demo.mybluemix.net/.

© Springer International Publishing Switzerland 2016
B. Davis et al. (Eds.): CNL 2016, LNAI 9767, pp. 48–57, 2016.
DOI: 10.1007/978-3-319-41498-0_5

allow users to be both expressive like with natural languages and computable like programming languages. These characteristics provide the basis for asking how effective it is to speak a controlled natural language, as it appears similar enough to natural language to be spoken and also be processed by computers for further use. With the variety of online speech recognition services available, the imperative in developing such a speech interface is to use a free, portable service to match the portability of the existing web-based interface of the PENG^{ASP} system [5]. Accordingly, we have decided on using Google's Web Speech API due to its accessibility and easy integration with the Google Chrome web browser; apart from that Google's Web Speech API is the only one of the three above-mentioned services that is completely free of costs.

While there exists substantial research on the design of text-based interfaces for controlled natural languages [1–3, 8, 12, 13, 16], there exists basically no research that investigates how an existing controlled natural language can be augmented with a speech interface and integrated with an online speech recognition service. Kaljurand and Alumäe [7] designed and implemented several topic-specific controlled language grammars for speech recognition which take the properties of speech into consideration. Their work is similar to the work that has been done in the context of W3C's Speech Interface Framework [10] where syntactic (context-free) grammars [6] and semantic representations [17] for these grammars have been standardised for the use in speech recognition. Our work is different from these approaches, since we do not aim to write a controlled grammar from scratch for a speech recogniser but try to interface an existing controlled natural language with an online speech recognition service.

The rest of this paper is structured as follows: In Sect. 2, we briefly introduce the current architecture and text-based interface of the PENG^{ASP} system. In Sect. 3, we present the requirements to a speech interface for a controlled natural language processor, followed by the requirements to a Web speech service. In Sect. 4, we present the design and implementation of our speech interface and its integration with the existing PENG^{ASP} system. In Sect. 5, we discuss experiences of using the speech interface. In Sect. 6, we identify a few potential future research directions, and finally in Sect. 7, we present our conclusion.

2   Current Architecture of the PENG^{ASP} System

The PENG^{ASP} system is designed to allow users to write non-monotonic specifications in controlled natural language [5]. The system operates with a web-based predictive text editor that continually communicates with the controlled natural language processor to provide lookahead categories, anaphoric expressions and spelling suggestions to the user to guide the writing process of their specification.

The PENG^{ASP} system, as shown in Fig. 1, has a client-server architecture, with the predictive text editor being the client that communicates with the controlled natural language processor through an HTTP server; the controlled language processor uses an answer set programming tool [4] as reasoning service. The predictive text editor provides an accessible and portable web-based interface, as illustrated in Fig. 2. The lookahead categories available via a pull-down

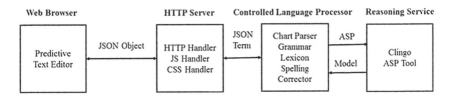

Fig. 1. Architecture of the PENG^{ASP} system

menu inform the user about the syntactically acceptable words at each position of the sentence and allow the user to complete the sentence by selecting the next desired approved word. Accessible anaphoric expressions are continuously updated during the writing process and available via another pull-down menu.

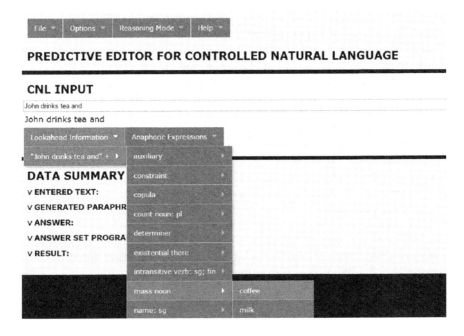

Fig. 2. Web-based predictive text editor of the PENG^{ASP} system

The server is implemented in SWI Prolog[4] and receives the words of the specification from the text editor via JSON objects[5]. The language processor then checks that the input conforms with the properties of the PENG^{ASP} language specified in an unification-based grammar, expands a discourse representation structure, and returns lookahead categories, anaphoric expressions and spelling

---

[4] http://www.swi-prolog.org/.
[5] http://www.json.org/.

suggestions to the client. After each new sentence, the discourse representation structure is translated into an executable answer set program [11] and this program is sent to the reasoning service [4] that tries to build a model for various reasoning tasks such as consistency checking and question answering [5].

3   Speech Interface and Web Speech API Requirements

In considering the development of a speech interface for the PENG^{ASP} system, multiple requirements were identified to shape the overall design of the speech interface. The requirements identified are as follows:

1. The speech interface must use the current PENG^{ASP} system framework in order to integrate easily. In particular, the existing PENG^{ASP} system should still be usable when not using the speech interface.
2. Google's Web Speech API must be used to implement the speech recognition functionality in the speech interface. This functionality must allow users to speak the controlled natural language PENG^{ASP}.
3. The interface must be designed to have high ease of use, particularly by tailoring existing functionality to suit the new mode of input.
4. The interface must allow alternate methods of input in order to reduce potential imprecision with the speech recognition functionality.

The functionality of Google's Web Speech API also shaped the requirements and design of the speech interface. The API was designed by Google as a JavaScript API for use in the Google Chrome web browser [15]. The speech recognition functionality is a multi-stage process that occurs through the Internet. Firstly, the client will activate the recording of voice data which is continually sent to Google's web service for recognition. The service will attempt to transcribe the words using speech recognition technology and send these results back to the client in form of a JSON object. The API acts as a black box with there being limited levels of customisability.

4   Interface Design and Implementation

The speech interface was designed to greatly integrate with the existing PENG^{ASP} architecture such that it is an interface to the existing system, rather than a separate system in itself. This was decided in order to benefit from the existing infrastructure while also allowing the speech interface to be used with future versions of the PENG^{ASP} system without requiring significant modifications. Additionally, the interface was designed to sit above the current system so that its implementation would not significantly alter the existing architecture. In line with these design factors, we included in the design the ability for users to switch between the existing interface (text mode) and the speech interface (speech mode) via an input mode selection menu.

A significant design choice involved developing the speech interface as a multimodal interface, as opposed to using only speech-based controls. This decision was made due to the inherent imprecision that speech-based controls are prone to. The speech recognition functionality was consequently provided as an alternative to text input, rather than designing the interface to be speech-only or speech-dominant. This design also addressed the privacy issues present in recording voice data to be sent over the Internet, because it gives the user control over when the speech recognition functionality is activated. The other ways of input to the interface are keyboard-based, for typing in the text field and submitting sentences, and mouse-based, for activating speech recognition and selecting options from menus.

An important aspect of the design of the interface was the need for simplicity due to the new mode of input using speech. The existing pull-down lookahead categories of the text-based interface provide extensive and fine-grained syntactic information and place emphasis on using the mouse to navigate through a hierarchy of menus. We decided to use a simpler menu structure by grouping some of the lookahead categories together and creating a fixed number of scrollable drop-down menus that are active or inactive during the writing process.

A design choice made to enhance the speech-related functionality of the interface was the collection of an n-best list of alternatives to the speech recognised words. This list is sent with the respective token to the controlled language processor and used for potential spelling suggestions in situations where a similar sounding word was transcribed instead of the intended word. If a token is not in the lexicon, then the spelling corrector of the language processor calculates the edit distances for that token and its alternatives using the relevant lexical entries and returns a ranked list of spelling suggestions. Note that this process is syntax-sensitive and only syntactically correct suggestions are returned and displayed. The extended architecture of the speech-enabled $PENG^{ASP}$ system is as shown in Fig. 3.

We implemented the speech interface primarily through JavaScript (ECMAScript[6]), JQuery[7] and HTML[8] for the web-based client, while some changes were made to the SWI Prolog server. The speech recognition functionality was implemented as an alternative to text input that is activated through a microphone button next to the text field. Due to the limited customisability of the speech API, we implemented text manipulation on the speech recognition results to standardise the transcription process. The main aspect of this was to truncate the transcription result into a single token in order to follow the token-based processing of the $PENG^{ASP}$ system. This requires users to speak each word distinctly so that the relevant lookahead categories and spelling suggestions can be generated for each word in the sentence. For example, if the transcribed result was "John *something else*", only the token "John" would be added to the text field and each subsequent word would need to be spoken separately. This approach was chosen because it avoids the risk of incorrect transcriptions in the middle of a sentence, which would be more difficult to correct given the token-based approach of the $PENG^{ASP}$ system.

---

[6] http://www.ecma-international.org/publications/standards/Ecma-262.htm.
[7] http://api.jquery.com/.
[8] https://www.w3.org/TR/html5/.

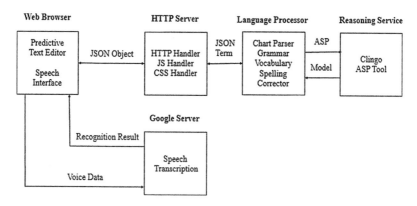

Fig. 3. Extended architecture of the speech-enabled PENG^{ASP} system

To complement the speech recognition functionality, we added buttons to provide basic control over the input transcribed to the input field. The undo button allows users to remove the last token in the text field, while the clear button removes all tokens in the text field. In addition to this, we added an updating ellipsis below the text field that appears when the speech recognition service has detected a word and is in the process of transcribing the word, which provides users with feedback on the speech recognition process.

The simplification of the lookahead categories was implemented on the server side, which involved checking the categories of the generated lookahead words and condensing these into a smaller number of categories to send back in the JSON object. This implementation greatly simplified the process of locating the correct lookahead words and is more accommodating to users with less of a formal understanding of English.

In addition to this, the problem with the pull-down menus was addressed by adding drop-down menus for the new lookahead categories. Drop-down menus were selected as they provided easy access to the list of words while also avoiding flickering as the categories update. As well as this, the labels of the categories change colour to represent whether there are syntactically correct words in that category for the next position in the sentence. The labels turn green when the category is active and red when the category is inactive. Through the reduction in the number of categories, the position of the drop-down menus on screen at all times and the colour notification of labels, the lookahead categories are now more suitable to speech input.

Similarly, we implemented the spelling suggestion functionality with a drop-down menu that appears once a spelling mistake has been detected. When the server sends data for spelling suggestions, a drop-down menu populated with these choices appears and stops the speech recognition functionality. This was done to prevent the user continuing to add words after a spelling mistake and then having to backtrack to the incorrect word. This implementation requires the user to correct the spelling mistake before they can continue speaking.

To implement the design of the speech interface as an extension of the existing web-based text editor, we added an input mode selection menu at the top of the interface. This approach preserved the existing interface to allow users to receive the extensive lookahead categories and avoid the speech-related functionality. For the n-best list of alternatives, we used an associative array to store and retrieve the results from. This data structure was chosen as there were frequent issues trying to implement it using a 2-dimensional array since the non-speech methods of input did not create n-best lists and would have required a complex approach to function correctly. A stack-based approach was considered, but this would have required extensive modifications to the existing methods for sending the tokens to the server. The associative array approach provided a simple solution and was particularly useful for words that were repeatedly spoken. After these features were implemented, the speech interface appears as shown in Fig. 4 below.

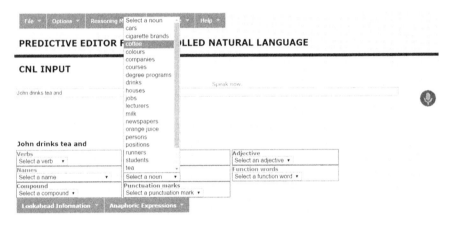

Fig. 4. Speech interface for the PENG^{ASP} system

5 Discussion

In developing and using the speech interface for the purpose of speaking the controlled natural language PENG^{ASP}, we have found the interface to generally have positive qualities as well as some negative qualities. It is worth highlighting both of them.

We found the API to be intuitive from both a development and end-user perspective due to its clear specification that greatly assisted with integrating it into a web application. The ability to integrate it easily in development allowed its actual use to be logical and consistent with the predictive text editor interface as a whole. Additionally, the speech recognition results are reasonably accurate and responsive, which is a good outcome for a free-to-use service.

However, the API did provide some issues in development and use, which can be expected for a free-to-use service. Firstly, there is a limited amount of customisation available for the speech recognition process, so this requires more extensive client-side processing for functionality that the API does not perform. This was evident in the need to implement text manipulation for the recognition results, since the results were returned as blocks of text reflecting the pace at which the user spoke rather than as separate tokens for each word. The API also has occasional problems with the detection and transcription of speech, which slows down the speed of input and can be frustrating for the user. A particularly significant issue with the API is it being removed from the problem domain, which results in words outside of PENG^{ASP}'s lexicon being transcribed and requiring correction. This results in the system having to compensate for these differences and introduces inefficiencies in the system.

In designing the interface to use speech as an alternative to text input, we found that the speech recognition functionality integrated relatively well with the existing PENG^{ASP} system interface. This was in part due to the deliberate choice to avoid making significant changes to the existing interface and resulted in the speech interface maintaining the existing characteristics of the interface as a token-based predictive text editor rather than a speech-driven interface. Once the users have familiarised themselves with the speech interface and the PENG^{ASP} language, the process of speaking PENG^{ASP} becomes fairly intuitive. The syntax of the language is familiar enough to that of the English language to be able to speak it, and the lookahead functionality is helpful when considering what word could be spoken next.

The most evident issue in speaking the controlled language PENG^{ASP} is the pace at which it is spoken compared to that of the English language. Since the speech recognition process in the speech interface requires each word to be spoken individually, it provides an artificial quality to speaking and limits the ability of PENG^{ASP} to be perceived as a true spoken language. In addition to this, there is a level of system learning and experience required in order to have the speech recognition process function optimally.

The adaptation of existing functionality to reflect the new mode of speech input was a useful aspect of the speech interface. The drop-down menus for the lookahead categories greatly simplified the existing pull-down menu approach and provide useful information on which categories are available for the next word while speaking. Similarly, the spelling suggestion drop-down menu provided a visual cue for correcting words that had been added and prevents the need to backtrack multiple words to correct a mistake. Finally, the undo and clear buttons provide good levels of control over the speech input and minimises the need to perform extensive backtracking with the keyboard. Overall, the process of speaking the PENG^{ASP} language and the predictive features adapted to speech input provided an effective demonstration of the expressiveness of the language and the predictive suggestion interface of the PENG^{ASP} system.

6   Future Research

The result of the development of this speech interface has provided a few areas for further research on the topic of speech interfaces for controlled natural language processing. One area for research is to determine how to best define new words in the lexicon, particularly whether the speech mode can provide an effective implementation of this or if this functionality should be left to the text mode as it is currently the case. Another area for research is assessing whether any alternative speech recognition services could be used to enhance the performance of the interface while also remaining portable. The upcoming Google Cloud Speech API[9] that will be based on powerful neural network models looks like an interesting alternative since speech accuracy of this service will improve as usage grows and the service will also be able to handle noisy audio. An important feature to consider would be the integration of the $PENG^{ASP}$ lexicon with the speech recognition service's lexicon to prevent incorrect words from being transcribed by training recognition models for different domains.

7   Conclusion

In this paper we presented the design and implementation of a speech interface to the $PENG^{ASP}$ system. This speech interface allows the controlled natural language $PENG^{ASP}$ to be spoken which provides an interesting and novel way of expressing a textual specification. The ability to speak $PENG^{ASP}$ serves as a positive demonstration of the expressive qualities of the language, which for the most part reflect the characteristics of the English language. As such, it is fairly intuitive to speak the controlled language $PENG^{ASP}$ from a language perspective and becomes more so once becoming familiar with the elements of the speech interface, the syntax of $PENG^{ASP}$ and using the lookahead categories.

The speed of input using online speech recognition is expectedly slower than the existing text input and reduces the ability of speaking $PENG^{ASP}$ at the same pace as a natural language. However, writing or speaking a textual specification that corresponds to a formal target language will probably never be a fast process since crafting such a specification requires careful reflection about the application domain. Finally, the design of any speech interface for a controlled natural language processor should preferably be multi-modal in order to resolve the imprecision inherent in speech recognition and provide users with flexible control over their input.

References

1. Bernstein, A., Kaufmann, E.: GINO - a guided input natural language ontology editor. In: Cruz, I., Decker, S., Allemang, D., Preist, C., Schwabe, D., Mika, P., Uschold, M., Aroyo, L.M. (eds.) ISWC 2006. LNCS, vol. 4273, pp. 144–157. Springer, Heidelberg (2006)
2. Franconi, E., Guagliardo, P., Trevisan, M., Tessaris, S.: Quelo: an ontology-driven query interface. In: Proceedings of the 24th International Workshop on Description Logics (DL 2011) (2011)

---

[9] https://cloud.google.com/speech/.

3. Fuchs, N.E., Kaljurand, K., Kuhn, T.: Attempto controlled English for knowledge representation. In: Baroglio, C., Bonatti, P.A., Małuszyński, J., Marchiori, M., Polleres, A., Schaffert, S. (eds.) Reasoning Web 2008. LNCS, vol. 5224, pp. 104–124. Springer, Heidelberg (2008)
4. Gebser, M., Kaminski, R., Kaufmann, B., Schaub, T.: Clingo = ASP + Control: Extended Report. In: CoRR, arXiv:1405.3694 (2014)
5. Guy, G., Schwitter, R.: The PENG^{ASP} system: architecture, language and authoring tool. J. Lang. Resour. Eval. 1–26 (2016)
6. Hunt, A., McGlashan, S.: Speech Recognition Grammar Specification Version 1.0, W3C Recommendation, 16 March 2004
7. Kaljurand, K., Alumäe, T.: Controlled natural language in speech recognition based user interfaces. In: Kuhn, T., Fuchs, N.E. (eds.) CNL 2012. LNCS, vol. 7427, pp. 79–94. Springer, Heidelberg (2012)
8. Kuhn, T.: AceWiki: a natural and expressive semantic wiki. In: Semantic Web User Interaction at CHI 2008: Exploring HCI Challenges, CEUR Workshop Proceedings (2008)
9. Kuhn, T.: A survey and classification of controlled natural languages. Comput. Linguist. **40**(1), 121–170 (2014)
10. Larson, J.A.: VoiceXML 2.0 and the W3C speech interface framework. In: IEEE Workshop on Automatic Speech Recognition and Understanding (ASRU 2001), pp. 5–8 (2001)
11. Lifschitz, V.: What is answer set programming? In: Proceedings of AAAI 2008, pp. 1594–1597 (2008)
12. Power, R.: OWL simplified English: a finite-state language for ontology editing. In: Kuhn, T., Fuchs, N.E. (eds.) CNL 2012. LNCS, vol. 7427, pp. 44–60. Springer, Heidelberg (2012)
13. Schwitter, R., Ljungberg, A., Hood, D.: ECOLE: a look-ahead editor for a controlled language. In: Proceedings of EAMT-CLAW03, Dublin, pp. 141–150 (2003)
14. Schwitter, R.: Controlled natural languages for knowledge representation. In: Proceedings of COLING 2010, Beijing, China, pp. 1113–1121 (2010)
15. Shires, G., Wennborg, H.: Web Speech API Specification, W3C Community, Final Report, 19 October 2012
16. Thompson, C.W., Pazandak, T., Tennant, H.R.: Talk to your semantic web. IEEE Internet Comput. **9**(6), 75–78 (2005)
17. Van Tichelen, L., Burke, D.: Semantic Interpretation for Speech Recognition (SISR) Version 1.0, W3C Recommendation, 5 April 2007

Asking Data in a Controlled Way with Ask Data Anything NQL

Alessandro Seganti[1]([✉]), Paweł Kapłański[1,2], Jesus David Nuñez Campo[1],
Krzysztof Cieśliński[1], Jerzy Koziołkiewicz[1], and Paweł Zarzycki[1]

[1] Cognitum, Wał Miedzierzynski 631, Warsaw, Poland
{a.seganti,p.kaplanski,j.campo,k.cieslinski,
j.koziolkiewicz,p.zarzycki}@cognitum.eu
[2] Gdansk University of Technology,
Gabriela Narutowicza 11/12, 80-233 Gdansk, Poland
http://www.cognitum.eu/

Abstract. While to collect data, it is necessary to store it, to understand its structure it is necessary to do data-mining. Business Intelligence (BI) enables us to make intelligent, data-driven decisions by the mean of a set of tools that allows the creation of a potentially unlimited number of machine-generated, data-driven reports, which are calculated by a machine as a response to queries specified by humans. Natural Query Languages (NQLs) allow one to dig into data with an intuitive human-machine dialogue. The current NQL-based systems main problems are the required prior learning phase for writing correct queries, understanding the linguistic coverage of the NQL and asking precise questions.

Results: We have developed an NQL as well as an entire Natural Language Interface Database (NLIDB) that supports the user with BI queries with minimized disadvantages, namely Ask Data Anything. The core part - NQL parser - is a hybrid of CNL and the pattern matching approach with a prior error repair phase. Equipped with reasoning capabilities due to the intensive use of semantic technologies, our hybrid approach allows one to use very simple, keyword-based (even erroneous) queries as well as complex CNL ones with the support of a predictive editor.

1 Introduction

The quality of modern decision-making processes is dependent on the ability to collect and analyze data. Well known approaches like e.g.: Relational Databases, Graph Databases or Online Analytical Processing (OLAP) Cubes [6], allow one to execute a query to an underlying database engine in a specific query language like e.g.: (correspondingly) SQL, SPARQL [10], and MDX [19]. OLAP Cubes are widely used in the industry and are considered the standard for defining the dimensions and access data for analytics. Still to create a query in any of the

---

Supplementary Information: Supplementary materials are available at Cognitum website: http://cognitum.eu/SmartBI.

© Springer International Publishing Switzerland 2016
B. Davis et al. (Eds.): CNL 2016, LNAI 9767, pp. 58–68, 2016.
DOI: 10.1007/978-3-319-41498-0_6

systems, the ability to use a complex query language and the knowledge about the structure of the underlying data is required. As a consequence, often decision makers cannot use directly these tools. There is thus a need for tools offering query-result loops, where the query is tailored within an interactive process that does not require any large prior learning and preparation. This way of querying data is supported by Natural Query Language (NQL).

The typical architecture of an NQL oriented application consists of three components: (1) an NQL-based user query interface that is also responsible for the transformation of a natural language query into a formal, machine-readable database query, (2) an underlying database system and (3) a textual or graphical reporting component that presents the results of database computations. We call such systems Natural Language Interface Databases (NLIDB).

In this paper we present an NQLIDB that we built: Ask Data Anything. This system is comprehensive of : a web application for interacting with data using natural language queries and a server application for processing the natural language query and executing it. We will in a first part highlight the main background to this article, then we will present the system's architecture. In a third part we will explain how ADA NQL queries are built and parsed. Finally we will evaluate our approach.

1.1 History of NLIDB

LUNAR [23] was the first NLIDB system that allowed natural language to be used to query a database about samples of moon rocks, however, nowadays this system is considered to be very limited in linguistic capabilities [18]. REN-DEZVOUS [4] was the system that implemented the "man-in-the-loop" way of human-database interaction based on a dialogue with a machine. Within the dialogue system it was possible to clarify all the difficulties found during the initial user input by helping the user to formulate queries. LADDER [11] was a general purpose NLIDB that was able to be connected to different underlying DBs, but at the same time it used grammars that were application-dependent making the system hardly portable. CHAT-80 [22] transformed English into Prolog expressions that were then evaluated against an existing database. CHAT-80 was a foundation for other experimental systems e.g.: MASQUE [1] and PRE-CISE [20]. ACE - Attempto Controlled English [8] is a Prolog-based, widely adopted general purpose language that allows a CNL-based NQL to be built. CNLs like ACE, being very precise and expressive require, at the same time, the use of a predictive editor that forms a kind of rails on which the user can write a syntactically correct sentence.

Data can often be represented as a graph of interconnected entities. In relational databases, relations are formally defined and expressed using table and database schema. On the other hand, semantic technologies allow Description Logic (DL) [3] to be applied with OWL Ontologies [12] on top of the given graph of resources (RDF graph [16]). Both approaches are used to keep and describe data semantics.

Also, modern NQLs are configurable with certain domain-specific ontologies, making the NQL core domain-agnostic. In the {AskMe*} system [17] an ontology is generated when the system is connected to a database. The generator processes the schema of a given database and generates an ontology that contains knowledge about the domain, properties, relationships and constraints that already exist in the given database. The ontology is then used to automatically generate a specific parser. Another example of a modern approach is SWSNL [9], where a natural language query is analysed by a linguistic component combining various Natural Language Preprocessing (NLP) technologies and translated into a SPARQL [10] query.

2   Ask Data Anything

2.1   The Ask Data Anything Architecture

Ask Data Anything (a NLIDB system developed by Cognitum) is a client-server application based on an MVC pattern with a web-browser based UI. The UI allows a NQL query to be entered and executed with the support of a predictive editor. The result-set of the query execution is presented on a wide range of reports including tables, charts and maps (Fig. 1).

Fig. 1. Ask Data Anything screen-shot from query execution result-set presented on a map

Ask Data Anything is made on top of the following core components (Fig. 2):

1. The Parser takes the input natural language query and translates it to a formal ADA NQL query by performing tokenization on it, fuzzy string matching and error correction.

2. The Engine first takes the formal ADA NQL query and extends it using the ontology elements then translates the query into a database specific query (SQL,...), finallly it sends the query to the database. The results are sent back to the UI.
3. The Database must implement the specific interface. It can be either an SQL database, a Graph database, a OLAP cube or a CSV file.
4. The Ontology Management System contains domain specific knowledge that extends the query execution process.
5. The ADA NQL support component is used in the predictive editor for auto-completing the user query in real time. It takes the partial ADA NQL query currently entered by the user, and tries to match the query to a known pattern using specific heuristics based on the underlying CNL. Then, hints relative to the recognized ADA NQL pattern are added and sent back to the UI as a list of words to complete the query allowing the user to select the most appropriate one.

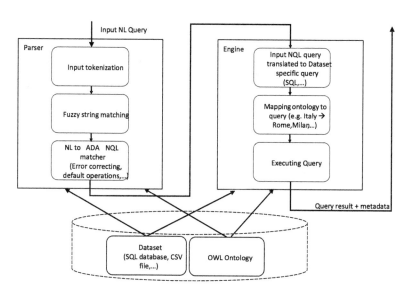

Fig. 2. ADA NQL query execution

2.2   The Ask Data Anything Parser

The Parsing Algorithm. The algorithm for parsing the natural language query we get from the user to the formal ADA NQL query expected from the underlying engine uses various steps and techniques. The steps in the parser are currently:

1. tokenize the input query and stem the verbs.
2. tag each token (e.g. recognize if it is a dimension, a keyword, a number, a datetime,...).

3. for each tagged token, try to match one or more token (e.g. constraint, operation, subsetting, aggregation,...) to build ADA CNL expressions (see Sect. 3).
4. check that all necessary pieces are inside the query, if not throw an error.

In Steps 2 and 3, we match the input string to the dictionary of possible words if the token is not recognized (e.g. Prodtcs is matched to Product). The pattern matching works by applying several string comparison algorithms and by creating a score for each element in the dictionary.

In Step 3, we are prioritizing the match between the various operation categories we support; the order is: subsetting, constraints, operations and aggregation. Each one of the searched operation categories is matched using chart parsing (similar to the Earley parser [7]). The formal ADA NQL query resulting from the parsing is then translated in the engine into a query to the Knowledge base (using the OCNL interface to SPARQL [14]) and a query to the Database (SQL,...).

3   ADA NQL

The Ask Data Anything natural query language (ADA NQL) is a proprietary query language composed of two different CNLs: a simplified version of the Ontorion Contolled Natural Language (OCNL, [13,24]) and the ADA CNL. Thus we can say that the ADA NQL queries are composed of one controlled natural language (the OCNL) embedded into another CNL (the ADA CNL); the idea of embedded controlled languages is presented in [21] and corresponds to our approach. OCNL is a controlled natural language (similar to Attempto [8]) to express knowledge with an English like syntax; this language is fully compatible with W3C standards OWL, RDF and SWRL. The ADA CNL is a subset of the English language that can be translated to formal database queries. Both of the languages can be used at the same time in the Ask Data Anything interface. The EBNF grammar for the OCNL language can be downloaded here while the one for the ADA CNL here.

Table 1. Example of a typical input data for Ask Data Anything.

Product	City	Brand	Date	Price
P1	Rome	Nike	21/02/2015	22.3
P2	Madrid	H & M	03/12/2014	44.0
..	..	...	..	..

Throughout this section we will use a simple example to explain how the ADA NQL queries are constructed. Let's imagine having a dataset of Products in Cities with Brands and Prices. In this case, the dataset would have columns like: Product, City, Brand, ... (see Table 1) while in the OWL ontology there will be defined statements like: Italy contains Rome, France contains Paris, Rome is a city, Paris is a city, ... (the OWL ontology here is written using the OCNL language equivalent to OWL/RDF, see [13,24] for further details). The combination of the data and the ontologies is called a Semantic Data-Set. In this section we will also use the convention to show elements of the grammar in **bold** and NQL queries and CNL in *italic*.

Dimensions and Operations. The ADA CNL queries begins with an **Operation** specification (sum, average,...) followed by (possibly more than one) **Dimension** specification. The **Dimension** specification(s) is (are) the only required grammatical part(s), all others are optional.

Each **Dimension** is assigned with a type inferred by parsing a subset of the data together with the information modelled in the supporting ontology. Currently, the types supported by the ADA CNL language are: Numerical, Date-time and Text, for the types understood directly from the data and: Location/-Geolocation, Latitude and Longitude, Hierarchical (defined in the supporting ontology, dimension that have super concepts grouping the values (e.g. *american-brand* for the Brands column,...) and Row (defined in the supporting ontology and represent data from multiple columns in a single row). In the Example presented above, the types will be (ColumnName : Type): Product : Text, City : Location, Brand : Hierarchical, Date : Date-time, Price : Numerical. For an extended description of how Ask Data Anything is coupling the concepts with the data see [5].

Operations and types are matched in the parser to check that the query makes sense so for example the query *Sum Product* is not allowed because Product's type is Text. For the same reason, *Sum Price* is allowed.

Subsetting. The next part of an ADA CNL query defines the **Subsetting** e.g. the filters to apply to the data. There are two ways of defining the subsetting: by specifying an SQL-like filter (keyword *where*), by using the keyword *in* or by using the pattern from..to.. (for date time filters only).

In the first case the general syntax is "**where Dimension Relation** Data" where as for the operations, the **Relation** and the **Dimension** are matched by the dimension type. Thus for example the query *Product where Price > 44* or *Product where Product == "P1"* are correct while the query *Product where Brand > 22* is incorrect.

In the second case the syntax is **in Dimension** or **in** Instance (from the ontology) so for example: *Product in P1, Product in 2015* or *Price in Italy* are valid queries.

In the third case, typical queries will look like: *Product and Brand from year 2015 to/until year 2016.* In the from..to.. part, various date time strings are recognized (e.g. *from 1st of July 2015 to/until 23rd of October 2015* or *from 07/01/2015 12:23 to/until 08/02/2015 09:22*).

Aggregation. The fourth part is the **Aggregation** which allows data to be grouped in subsets. The syntax for aggregation is **by Dimension** or **by Location** or **by** time period (day, year,..) or **by** Concept from the ontology. Multiple aggregations are allowed (e.g. by *country* and by *day*). Some aggregations require operations and others do not (e.g. by day can be used with or without operations on the dimension, while by country needs an operation). Some example of queries using aggregations are *Sum Price by spanish-brand*, *Count Product by day and by country* or *Count Brand by City and by Product and by year*.

Embedded OCNL. The OCNL used in the previous queries where all references to instances (e.g. *Italy*) or concepts (e.g. *country, american-brand*) defined in the ontology. Still also full OCNL queries can be embedded into the ADA NQL. Thus queries like *Sum price in fish that has-size greater-than 10 and is not a frozen-product and-or is a freshwater-fish on a piechart* are valid ADA queries. This query contains the following OCNL expression: *fish that has-size greater-than 10 and is not a frozen-product and-or is a freshwater-fish* that evaluates into a complex SPARQL query that can be executed on the knowledge base.

Query-Result Loop. Our approach to query disambiguation follows one golden rule: try to match as much as possible, execute the query and then ask the user if s/he meant something else in a query-result loop. This kind of loop improves the user experience as the user is able to use trial and error to understand how the query language works. Still, we always show what is the understood (right side of Fig. 3) query under the query that the user has written. Furthermore, we use coloring and explanations to explain to the user why some part has been added or modified in the query. This query-result loop is quite effective but it needs to be noted that while the autocomplete is shown in real-time, the interpretation of the query together with what has been modified is shown only after parsing the full sentence.

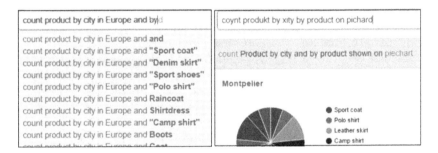

Fig. 3. ADA NQL supporting features: (left) predictive editor, (right) "entered" and "understood" query. (Color figure online)

The higher the complexity of the data, the more difficult it is to match the query of the user to the underlying data structure. The user can thus give us hints on which is the correct column to take. One case where disambiguation is needed is if the same string is present in more than one column, e.g. Nike is a brand and a brand name. In this case, we take one of the columns automatically and a warning is shown to the user giving him the possibility to perform the query again using the string from the other column. Currently, in the query language it is possible to specify the column (see the BNF for the ADA CNL) that should be taken by giving the value in parenthesis (e.g. if the brand Nike exists in two columns Brand and Brand Name then the user can specify Nike (Brand) or Nike (Brand Name)).

Another case in which disambiguation is needed is for datetime subsetting. So if the user writes *from March 2016 to July 2016* and there is more than one date column, then we take one of them by default. Still the user can tell us if he/she wants to subset the datetime constraint for another column by adding the name of the column at the end of the subsetting part (e.g. *from March 2016 to July 2016 for "Some Date Column"*).

4   Evaluation

4.1   General Advantages and Disadvantages of NQL and NLIDB

To evaluate our system, firstly we tried to place it in the spectrum of well-known advantages and disadvantages of NQL and NLIDB (see [2])

Advantages. Using ADA NQL does not require prior learning of a database query language like SQL or SPARQL. For example the ADA NQL query *Sum price by vendor in fish and in frozen-products* translates underneath in a complex SQL query containing a SPARQL query like:

```
select sum( price ) from dataset where
  product in (
  select distinct ?x {
    ?x rdf:type ns:fish .
    ?x rdf:type ns:frozenProduct
})
```

We argue in this article that the ADA NQL form of the query is more intuitive to build in a query-result loop.

Furthermore Ask Data Anything is fault-tolerant, providing the user with a result-set together with an information about what is "understood" as a query (right side of Fig. 3). For example the following erroneous Ask Data Anything query: *avrage prize incofee by municipality on map* is rewritten into: *average prize in coffee by municipality on a map*, prior being evaluated by the Ask Data Anything engine.

Disadvantages. The ADA NQL queries are quite intuitive to understand, still the query language needs to be learned. In order to mitigate this problem, we built a UI helping the user to choose the correct answer by autocompleting the next element of the query and by automatically correcting the formal errors in the query. Furthemore in the web application there is a visual way of exploring what is contained in the data together with the types inferred by ADA and other information.

Another problem of this kind of systems in general, is that a domain knowledge needs to be built in order for the knowledge base part to be used fully. In ADA, this problem is mitigated by the fact that the OCNL language is compatible with the standards thus giving the user the possibilty to reference already existing domain knowledges and use them inside ADA NQL queries. Furthermore we are currently building together with our clients basic knowledge for different industries (e.g. aeronautics, factory management,...) that can be used as a starting point.

4.2 PENS Classification Scheme

The ADA NQL is a CNL for databases queries and therefore it can be classified according to the well-known Classification of Controlled Natural Languages - PENS [15]. In terms of the PENS Classification Scheme ADA NQL has the following properties:

P^3 is restricted by the grammar, however, it is hardly formally defined as it allows for erroneous statements. Nevertheless, it gives the opportunity to be automatically interpretable even during autocompletion. According to PENS - it is a reliably interpretable language

E^2 is able to handle complex OCNL concept expressions. OCNL maps to OWL2 and ultimately to Description Logics, therefore ADA NQL is a language with low expressiveness

N^4 queries have the form of natural language, however, there are infrequent exceptions and unnatural means for clarification - it is a language with natural sentences

S^4 a description of the language requires more than one page but not more than ten pages - it is a language with a short description (see [5]).

We also recognise the following CNL properties in ADA NQL: **C** (the goal is comprehensibility); **F** (the goal is formal representation (including automatic execution)); **W** (the language is intended to be written); **S** (The language is intended to be spoken); **I** (The language originated from industry).

5   Conclusion and Future Work

In this paper we presented Ask Data Anything, an NQLIDB that we implemented. We showed that the system we have implemented is composed of a UI together with a natural query interface. We have presented the ADA NQL query language and showed how it can be used to make complex queries to data sets. As explained, the system is robust so that erroneous queries are automatically corrected (if possible). This approach, simplifies the interaction with the system in a query-result loop. Furthermore, as the ADA NQL embeds the OCNL controlled natural language for knowledge bases, the user can easily integrate ontologies written using W3C standard technologies into the system and use them to enhance the data.

ADA NQL is $P^3E^2N^4S^4$ class and in our opinion, reduces the common disadvantages of NQLs. We also would like to point out that the NQL language offers more than what presented here as for example: simple access to semantic tagging, synonyms and other features given by the ontology that could not be presented here for space costraints. For a full description of the NQL language, please check [5].

We are currently extending the query language to work with more than one table/file. In this case, we need to implement a disambiguation strategy for choosing which table the dimension is from. We are also planning on making an experiment to show how the interaction with Ask Data Anything simplifies and enhance the retrieval of data from a database.

6 Conflict of Interest

All authors work for Cognitum, the company behind the Ontorion Server, Fluent Editor and OCNL.

References

1. Androutsopoulos, I., Ritchie, G., Thanisch, P.: MASQUE/SQL - an efficient and portable natural language query interface for relational databases. Database technical paper, Department of AI, University of Edinburgh (1993)
2. Androutsopoulos, I., Ritchie, G.D., Thanisch, P.: Natural language interfaces to databases - an introduction. CoRR cmp-lg/9503016 (1995). http://arxiv.org/abs/cmp-lg/9503016
3. Baader, F., Calvanese, D., McGuinness, D., Nardi, D., Patel-Schneider, P.: The Description Logic Handbook: Theory, Implementation and Applications. Cambridge University Press, New York (2003)
4. Codd, E.F.: Seven steps to rendezvous with the casual user. In: IFIP Working Conference Data Base Management, pp. 179–200. iBM Research Report RJ 1333, San Jose, January 1974. http://dblp.uni-trier.de/db/conf/ds/dbm74.html#Codd74
5. Cognitum.: Askdataanything! 2015 documentation (2016). http://docs.cognitum.eu/AskDataAnything2015/. Accessed 13 May 2016
6. Datta, A., Thomas, H.: The cube data model: a conceptual model and algebra for on-line analytical processing in data warehouses. Decis. Support Syst. **27**(3), 289–301 (1999)
7. Earley, J.: An efficient context-free parsing algorithm. Commun. ACM **13**(2), 94–102 (1970). http://doi.acm.org/10.1145/362007.362035
8. Fuchs, N.E., Schwertel, U., Schwitter, R.: Attempto controlled English – not just another logic specification language. In: Flener, P. (ed.) LOPSTR 1998. LNCS, vol. 1559, pp. 1–20. Springer, Heidelberg (1999)
9. Habernal, I., Konopík, M.: SWSNL: Semantic web search using natural language. Expert Syst. Appl. **40**(9), 3649–3664 (2013). http://dblp.uni-trier.de/db/journals/eswa/eswa40.html#HabernalK13
10. Harris, S., Seaborne, A.: SPARQL 1.1 query language (2013). http://www.w3.org/TR/sparql11-query/. Accessed 21 Sept 2015

11. Hendrix, G.G., Sacerdoti, E.D., Sagalowicz, D., Slocum, J.: Developing a natural language interface to complex data. ACM Trans. Database Syst. **3**(2), 105–147 (1978). http://doi.acm.org/10.1145/320251.320253

12. Hitzler, P., Krötzsch, M., Parsia, B., Patel-Schneider, P.F., Rudolph, S.: OWL, web ontology language (2004). http://www.w3.org/TR/owl2-primer/. Accessed 21 Sept 2015

13. Kapłański, P.: Controlled english interface for knowledge bases. Stud. Informatica **32**(2A), 485–494 (2011)

14. Kapłański, P., Weichbroth, P.: Cognitum ontorion: knowledge representation and reasoning system. In: Position Papers of the 2015 Federated Conference on Computer Science and Information Systems, FedCSIS 2015, Lódz, Poland, 13–16 September 2015, pp. 177–184 (2015). http://dx.doi.org/10.15439/2015F17

15. Kuhn, T.: A survey and classification of controlled natural languages. Comput. Linguist. **40**(1), 121–170 (2014). http://www.mitpressjournals.org/doi/abs/10.1162/COLI_a_00168

16. Lassila, O., Swick, R.R.: Resource description framework (RDF) model and syntax specification. W3c recommendation. W3C, February 1999. http://www.w3.org/TR/1999/REC-rdf-syntax-19990222/

17. Llopis, M., Ferrández, A.: How to make a natural language interface to query databases accessible to everyone: an example. Comput. Stand. Interfaces **35**(5), 470–481 (2013). http://dblp.uni-trier.de/db/journals/csi/csi35.html#LlopisF13

18. Nihalani, M.N., Silakari, S., Motwani, M.: Natural language interface for database: a brief review. IJCSI Int. J. Comput. Sci. Issues **8**(2), 600–608 (2011)

19. Nolan, C.: Manipulate and query olap data using adomd and multidimensional expressions. Microsoft Syst. J. **63**, 51–59 (1999)

20. Popescu, A.M., Armanasu, A., Etzioni, O., Ko, D., Yates, A.: Modern natural language interfaces to databases: composing statistical parsing with semantic tractability. In: Proceedings of the 20th International Conference on Computational Linguistics, COLING 2004. Association for Computational Linguistics, Stroudsburg (2004). http://dx.doi.org/10.3115/1220355.1220376

21. Ranta, A.: Embedded controlled languages. In: Davis, B., Kaljurand, K., Kuhn, T. (eds.) CNL 2014. LNCS, vol. 8625, pp. 1–7. Springer, Heidelberg (2014). http://dx.doi.org/10.1007/978-3-319-10223-8_1

22. Warren, D.H.D., Pereira, F.C.N.: An efficient easily adaptable system for interpreting natural language queries. Comput. Linguist. **8**(3–4), 110–122 (1982). http://dl.acm.org/citation.cfm?id=972942.972944

23. Woods, W., Kaplan, R., Nash-Webber, B.: The Lunar Sciences Natural Language Information System: Final Report. BBN report, Bolt Beranek and Newman (1972). https://books.google.pl/books?id=RhuEMwEACAAJ

24. Wroblewska, A., Kaplanski, P., Zarzycki, P., Lugowska, I.: Semantic rules representation in controlled natural language in FluentEditor. In: 2013 The 6th International Conference on Human System Interaction (HSI), pp. 90–96. IEEE (2013)

To What Extent Does Text Simplification Entail a More Optimized Comprehension in Human-Oriented CNLs?

Nataly Jahchan[1,2(✉)], Anne Condamines[1],
and Emmanuelle Cannesson[2]

[1] CLLE, University of Toulouse, CNRS, Toulouse, France
anne.condamines@univ-tlse2.fr
[2] Airbus Operations SAS, Toulouse, France
{nataly.jahchan, emmanuelle.cannesson}@airbus.com

Abstract. The aim of this study is to develop a new cockpit controlled language for future Airbus aircraft by using psycholinguistic testing to optimize pilot comprehension. Pilots are aided by cockpit messages in order to deal with different situations during aircraft operations. The current controlled languages used on the Airbus aircraft have been carefully constructed to avoid ambiguity, inaccuracy, inconsistency, and inadequacy [21] in order to ensure the safety of the navigation, operational needs, and the adaptability of the human-computer interaction to different situations in the cockpit. However, this controlled language has several limitations, mostly due to small screen sizes (limited number of words and sentences) and is highly codified (non-conforming to natural language syntax, color-coded, etc.) so that it requires prior pilot training in order to achieve fluency. As future cockpit design is under construction, we might be looking at a different flexibility margin: less limitations, different screen sizes, less coding, etc.

Keywords: Comprehension-oriented CNL · Controlled language · Airbus cockpit alarms · Human factors · Psycholinguistics · Text comprehension · Comprehension optimization

1 Introduction

Going back to the origins of controlled languages, we would find that the main goal of the first CNLs was to facilitate communication among humans – such is the case with BASIC English, created by Charles Ogden in 1930.

Readability research and controlled language production (BASIC English, PLAIN English, AECMA SE, etc.) have constantly been criticized for lack of empirical research that justify their rules and existence [8, 9, 13, 19]. Rudolf Flesch in his article "How Basic is Basic English?" [9] claims that Basic English "is neither basic nor English" and starts off with an example *"If I were Mr. Churchill, I would not like being reduced to calling Hitler* "a very bad man" *or a bomber* "an air plane sending down hollow balls full of substance with a tendency to go off with a loud noise", in reference

B. Davis et al. (Eds.): CNL 2016, LNAI 9767, pp. 69–80, 2016.
DOI: 10.1007/978-3-319-41498-0_7

to Basic English's arbitrarily selected 850 word vocabulary. He criticizes Ogden for *"deliberately avoid[ing] the scientific approach and not [being] lucky enough to find the key to simplicity by accident"*.

With time, the very nature of our modern day communication has pushed researchers to find different usages (i.e. other than CNLs for facilitating communication, mutual comprehension and ease of use) for controlled/simplified/processable/etc. languages such as automatic translation or formal notations in different domains (Industry, Academia, Government, etc.). See Kuhn [16] for a complete survey of the available CNLs and their usage.

AECMA SE, more recently known as ASD-STE (Aerospace and Defense Simplified Technical English), one of the most complete, widely used comprehension-oriented controlled languages (a language that has survived the test of time and is still used in the Aircraft maintenance domain, and across different aircraft manufacturers) was also accused of harboring anecdotal, intuition-based evidence to justify its many rules of use. The lack of scientific evidence was jarring at the time of AECMA SE's adoption in aircraft maintenance, specifically in aeronautics where the consequences of inaccurate comprehension could lead to potentially dangerous situations. According to Hinson [13], *"AECMA's Simplified English claims to be founded on readability research. It would be interesting to establish the nature, validity, and appropriateness of the research used. It would also be helpful to know of any research carried out on Simplified English manuals in use."*

To this effect, a wave of research studies in the mid-90's [3, 4, 8, 19, 22] was launched to acquire the much needed empirical evidence that AECMA SE lacked. These studies will be of great interest in our research. The experiments conducted and relevant results will be elucidated in Sect. 4.2.

2 Context

For the purposes of our research, we will only be interested in comprehension-oriented CNLs, that is to say CNLs whose main goal is to enhance/improve/optimize human comprehension in a given corpus. By improving comprehension we are targeting three main aspects: faster comprehension, more accurate comprehension, and limited training needs.

This paper is part of the author's PhD research launched by the Human Factors department of Airbus Operations SAS in Toulouse, France in collaboration with CLLE (Cognition, Langues, Langage, Ergonomie) laboratory of Toulouse 2 University. Based in Toulouse, an aerospace hub, CLLE laboratory has cultivated a knowledge base in the CNL domain and specialized corpora related to space and aviation (see [5, 17]. As such, the main goal of this study is to develop a new cockpit controlled language for future Airbus aircraft by using psycholinguistic testing to optimize pilot comprehension. Pilots are aided by cockpit messages in order to deal with different situations during aircraft operations. The current controlled languages used on the aircraft have been carefully constructed to avoid ambiguity, inaccuracy, inconsistency, and inadequacy [21] in order to ensure the safety of the navigation, operational needs, and the adaptability of the human-computer interaction to different situations in the cockpit.

However, this controlled language has several limitations, mostly due to small screen sizes (limited number of words and sentences) and is highly codified (non-conforming to natural language syntax, highly abbreviated, typographically variable, color-coded and so on (*cf.* Fig. 1)) so that it requires prior pilot training in order to achieve fluency. As future cockpit design is under construction, we might be looking at a different flexibility margin: less limitations, different screen sizes, less coding etc. Figure 1 is an example of different messages found at different locations in one of the corpora at hand (this is not an *exact*[1] replica of an alarm).

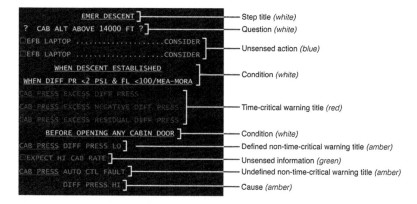

Fig. 1. Example of different messages in cockpit corpus (Color figure online)

Our corpus' controlled language is highly context specific, color-coded, contains a lot of technical jargon, and addresses an expert target audience, all of which make it hard to decipher to the layperson outside of the aeronautical domain. However, while very coded and seemingly unnatural, it nevertheless falls in the domain of CNLs, as it is engineered to meet a special purpose and uses specific rules of grammar and a set of restricted lexicon. Contrary to Kittredge [14] and Fuchs et al. [11], Kuhn [16] denotes that controlled languages are not necessarily always proper subsets of the underlying natural language because there can be small deviations from natural grammar and semantics, in addition to some unnatural elements like colors that are meant to increase readability. *"The subset relation is clearly too strict to cover a large part of the languages commonly called CNL"* [16].

3   Aim

"Natural language being such a breeding ground for ambiguity, to communicate just one set of meanings while excluding many others is often impossible" [6]. In a sense, that statement is true since natural language has theoretically infinite possibilities of

---

[1] For confidentiality reasons complete Airbus alarms cannot be released here. The lines in Fig. 1 are assembled from different alarms, and they are representative of the various types of information in the corpora.

expression and interpretation, but in another sense, natural language is the most common and constant tool in our cognitive process of everyday life. Syntactic constructions, morphological derivations, way of thought, all come naturally in the way we acquire them at an early age; or the way non-native speakers of a given language, let's say English, first learn the language at its most basic form and construction: naturally, without any control. Bisseret [2] regards natural language as a *"universal tool of representation and of thought communication."*

We therefore hypothesize that the exposure to natural language for both native and non-native speakers influences the way people will understand a certain text and respond to it efficiently. In other words, an unambiguous text written in a natural language construction would, in our opinion, be more easily understood than a coded, controlled, and syntactically non-conforming to natural language CL. This is due to speakers being more exposed to a certain natural language and its constructions in their usage of this language in their everyday life. Or so we hypothesize. We will endeavor in this study to find empirical proof to substantiate or deny this argument.

The idea is not to eliminate controlled language altogether. For then, without rules, common linguistic ambiguities would be very easy to come by. The real question is: what is the right balance? Researches have been quite adamant that "simplification" was the right way to proceed to achieve better comprehension. Readability, text-complexity, text-cohesion research have all focused on the process of simplification/controllability/structuration ([7, 18, 23, 24] among others). According to Van Oosten et al. [24], *"the concept of readability has been defined in a wide variety of ways, typically dependent on the author's intentions. For instance, Staphorsius (1994) defines readability of a text as the reading proficiency that is needed for text comprehension. The author's intention of designing a formula to determine the suitability of reading material given a certain reading proficiency is not without its influence in that definition. McLaughlin (1974), the author of the influential SMOG formula, on the other hand, defines readability as the characteristic of a text that makes readers willing to read on."* Or according to DuBay's 2004 definition *"what makes texts easier to read than others"."* In our case, readability is not about ease of reading, or reading proficiency or the characteristics that make readers willing to carry on reading. Readability in our sense is about usability of the text. What are the inherent qualities of a sentence that make it comprehensible? By comprehension, we mean that the information we want to transmit has been fully understood, the consequences of which should be the correct reaction to the information and the writer's intended meaning in the most optimal manner (fast and accurate comprehension and reaction).

In some CLs, simplification reduced the sentential elements to the basic essentials, and diminished the scope and complexity to the detriment of information loss. The following is an example of PLAIN English CL (controlled languages) taken from their website[2]:

A. High-quality learning environments are a necessary precondition for facilitation and enhancement of the on-going learning process.
B. Children need good schools if they are to learn properly.

---

[2] http://www.plainenglish.co.uk/.

According to the Plain English approach, these two sentences are synonymous, with sentence A being more difficult than sentence B. While that might very well be the case, sentence B does not say everything sentence A intends to say. The semantic field has been highly restricted. For instance, "learning environments" are not strictly limited to "schools", and not universities or home-schooling, tutoring etc. "Facilitation and enhancement" is not accurately summarized by "learning properly". The idea of an "on-going process" has been completely eliminated. In our opinion, those two sentences are in very little ways synonymous. Simplification has led to a substantial change/reduction of meaning that unless it specifically intended to do so, has failed to accurately "simplify". In other cases, making information more explicit and redundant caused the readers to lose what could be otherwise valuable time [3, 4, 8, 19, 22].

Codifying and abridging languages, controlling and simplifying, whether by using syntactic or other forms of ellipses could make a language difficult to assess for a lay speaker of a given language. That is to say, a codified language might require prior training and possibly more effort on the end user regarding direct and easy comprehension; a process that might well be exacerbated in situations of stress or danger. Therefore, the usefulness and usability of an acquired (in the sense of requiring prior learning) controlled language must be put to the test and undergo psycholinguistic scrutiny.

Our experimentation plan is to go against the tide of common comprehension-oriented CNL construction, in the sense that we will not be taking natural language and simplifying it, but rather taking a highly controlled codified language (therefore theoretically most simple) and "complexifying" it (bring it closer to natural language: theoretically most complex) in order to make it more accessible. In other words, we want to bring it back to a more natural state: give it a more natural language structure, syntactically and otherwise.Therefore, we are going backwards, towards natural language, while making sure not to fall in the trap of ambiguity.

Simplification does not necessarily have to start from an unsimplified text; such is the case with more formal representation languages such as Attempto Controlled English [10] that start with basic logical relations and gradually add complexity. In these cases, simplification is applied to force writers to write in a simple manner from the start.

In our case, we will go from a codified corpus to a more natural one[3], by using research that has been done on readability and complexity and test, bit by bit, how we can add sentential elements that would make the language closer to natural language structure of English. At the same time, by adding a sentence structure we would be limiting the different possible interpretations, therefore avoiding, as much as possible, elliptical ambiguities. This general thought has led us to delve into the different times research has empirically proven that controlling a language actually improved global human comprehension. And more than that, how much control was needed to actually achieve better comprehension, and what are the limits that could potentially render this control or oversimplification unsatisfactory/counter-productive?

---

[3] One main reason for the corpus being our starting point is that we are doing research on an applicative basis. As such, we must bear in mind that our end users rely on specific corpora and functions, and we have to take into consideration the potential evolution of their learning process rather than enforcing radical change.

4 State of the Art

4.1 Simpson and Hart

Carol Simpson [20] studied the effects of linguistic redundancy on pilot's comprehension of synthesized speech (a study done for Human Factors research in aviation in a psycholinguistics context in NASA's Ames Research Center). She showed that by taking the time to form clear unambiguous sentences using the same original keywords, the message was detected more accurately and pilot's reaction times was faster. For instance, the message "fuel low" was inserted in a sentence in the same order "The fuel pressure is low" and "gear down" was inserted in "The landing gear is down". The same goes for "Autopilot disengaged" and "The autopilot is disengaged". Response times to sentences were approximately 1 s shorter than response times to two-word messages. The results take into account the duration of the messages. That is to say, even though the duration of the stimuli containing the keywords in sentences was longer, the reaction times in total were still faster in the case of keywords in sentences than in the case of simple keyword messages. The experiment also showed that key words in sentences were approximately 20 percent more intelligible than key words presented alone.

Moreover, sentence-length messages appeared to require less attention to comprehend than two-word messages (Sandra Hart [12], concurrent study for NASA).

Cockpit alarms have a tendency to be presented in the form of short keyword messages rather than in the form of long sentences. Brevity is usually preferred because of the small window of time that the pilots have to react in time-critical situations. Therefore, the obvious way to economize on the time of stimuli presentation was to make the messages as short, precise, and unambiguous as possible so as to keep only the relevant information, and eliminate redundancy provided by a sentence structure, i.e. the suppression of syntactic sentential elements.

It was concluded in Simpson's research that the syntactic and semantic constraints provided by a sentence frame (which adds redundancy and explicitness) reduced the possible interpretations of keyword alerts. Furthermore, the pilot participants mentioned that *"the longer pattern of the sentence with extra words between the critical ones gives you more time to understand the words"* and in their case react faster to the alert. While these results are based on aural alerts, one could hypothesize that the same argument would work on written alerts.

Simpson's and Hart's concurrent studies offered the necessary background to start "de-codifying" our corpus by going towards natural language sentential structure. Contrary to the other human-oriented CLs' evaluations [3, 4, 8, 19, 22, 24], Simpson's study results showed that certain language structures (non-simplified natural language structures) actually decreased response time, which is a factor that is particularly of interest to us for optimizing comprehension. Additionally, to the best of our knowledge, it is one of the only experiments that tested accuracy of comprehension and time in short injunctive sequential messages as opposed to long chunks of text ([15, 18] among others).

4.2   AECMA SE Evaluations

We will now summarize the research experiments done on AECMA SE [3, 4, 8, 19, 22] which were done to substantiate SE's efficiency. These studies are important to us mainly because by questioning the extent of efficiency of controlled languages, we also question, to a certain extent, the legitimacy of controlling/(over)simplifying a natural language.

The researchers were interested in testing the effects of SE on comprehension, location of information on maintenance work cards and response time. They compared pre-SE work cards and their SE version. The experiment designs and independent variables differed between those 5 experiments: both native and non-native speaker participants or exclusively non-natives, in an English speaking country or not, using technicians or students, using easy vs. difficult work cards, using reading comprehension for testing or actual performance of maintenance, testing for subjects' reading comprehension and level of English or not, testing or not the work cards' text complexity using Flesch-Kincaid readability tests[4]. Some results show that while SE appeared to be significantly better for comprehension on the whole, it also came short of significance in several different conditions. As seen in the comparative table (Table 1), Shubert et al. [19] and Chervak et al. [4] are the only two studies that showed general SE superiority significantly. In Chervak [3], Eckert [8], and Stewart's [22] studies there were no significant results to substantiate SE superiority over non-SE versions. Furthermore, in Chervak et al. [4] and Shubert et al. [19] (the only experiments showing general SE significance) there was a significant interaction of comprehension of SE and non-SE by document type: The easy work cards (ones that described short and easy procedures as opposed to long and difficult ones) did not show any comprehension significance for SE, and only the hard ones did. Therefore, content is not significantly more comprehensible or easy to locate for the subjects working with the shorter easier procedure. Chervak et al. [4] showed that only certain work card types showed significant SE superiority over non-SE, which suggests that SE superiority, is document specific.

Finally, none of the experiments showed that SE significantly improved time. Shubert et al. [19] even noted that in the easier work cards the subjects reading SE documents required more time to respond. Stewart [22] also notes that participants in the SE condition needed to have a higher mean English-reading ability to obtain a mean task card similar to non-SE condition.

All of these studies concluded that while the superiority of SE seemed to be very document and condition specific, it did not adversely affect comprehension in the other conditions. Therefore, Chervak et al. [4] concluded that SE was suitable for use especially where it is needed most: in hard and long work cards and for non-native speakers. However, most interestingly Eckert [8] and Stewart [22] who only tested non-native speakers did not find any SE significance. Temnikova's [23] experiment is different from the other experiments since it was done 15 years later, and was testing a

---

[4] Readability tests designed to indicate how difficult a reading passage in English is to comprehend. They rely on measuring word length and sentence length to provide a grade level of the text or a reading ease level.

different CL: Controlled Language for Crisis Management (CLCM). It is relevant here because, like the previously mentioned 5 studies, it also tests a human-oriented CL psycholinguistically in a behavioral experimental protocol. Results showed that there was no statistically significant global superiority of the simplified CLCM over the "complex" natural language. It was significant in certain sets of text (again, document-specific) and it did not show any significance with regards to response time. All of these results are summarized in Table 1.

Table 1. Comparative table summarizing most relevant results of different CL evaluations.

Author/year	Shubert et al. 1996	Chervak et al. 1996	Chervak 1996	Eckert 1997	Stewart 1998	Temnikova 2012
Native and non-native	Both	Both	Native	Non-native	Non-native	Both
Participants: natives	90 natives	157 natives	18 natives	0 natives	0 natives	22 natives
Participants: Non-natives	31 non-natives	18 non-natives	0 non-natives	148 non-natives	41 non-natives (21 different countries)	83 non-natives
Profession	Engineering students	AMT's	9 maintenance students and 9 experienced mechanics	Aviation maintenance students	Electronics technician students	All walks of life(because not testing SE, but CLCM)
Country	English speaking	English speaking	English speaking	Non-English speaking (Mexico)	English speaking	N/A (Online experiment)
Procedure	Reading comprehension, between subject	Reading comprehension, between subject	Performing maintenance, between subject	Reading comprehension, between subject	Reading comprehension, between subject	Reading comprehension, between subject
Tested for English comprehension	No	Yes (but not specifically for non-natives)	No	Yes	Yes	No (only self-evaluation and not used in analysis)
General SE Significance: doc type	Yes	Yes	No (means followed trend)	No (means followed trend)	No (means followed trend)*	No (means followed trend)
Significance SE comprehension: easy	No	No	N/A	N/A	N/A (only 1 workcard)	N/A
Significance SE comprehension: difficult	Yes	Yes	N/A	N/A	N/A (only 1 workcard)	N/A
Significance: time/SE	No**	No (will not adversely affect)	No	N/A	No	No
Significance: time/native speaker	N/A	Yes (normal)	N/A	N/A	No	N/A
Significance type of workcards	N/A (only easy/difficult was tested)	Yes (only certain workcards)	N/A	N/A	N/A	Yes (only certain sets of text)

*The study also concluded that the SE participants required higher mean English-reading ability to obtain a mean task card test score similar to the non-SE participants.

What is interesting, however, is that for procedure B (easier) the subjects reading **SE versions of that document took slightly **longer** than those reading the Non-SE versions.

The results of these experiments are relevant to our study for two main reasons: 1- We are interested in time optimization and these AECMA SE and CLCM evaluations show that simplifying a language does not necessarily economize time and 2- because our corpus is made of short relatively uncomplicated sequential procedures and these results do not show CL superiority when it comes to easy procedures.

These last results and Simpson and Hart's research has led us to give a more concrete form to our hypothesis, that native language syntactic structure on a sentential level would help us optimize comprehension.

5 Approach

This paper being a preliminary approach to evaluating our proposed hypotheses based on the previously presented state of the art, we have not yet set the experiment in motion at this time. Our experimental protocol is still under construction and the variables are not yet fully defined. In this section, we will only evoke the general guidelines that will form our upcoming evaluations, and hope to have preliminary results ready for CNL workshop in July 2016.

5.1 Hypothesis

Coded format keywords inserted in natural language structure sentences have a faster reaction time and are understood more accurately (less mistakes in performance) than keywords in isolation (*cf.* Example 1).

EFB Laptop...................Consider *(Actual coded format)*
VS.
Consider using EFB Laptop. *(Proposed natural format)*

Example 1. Example of an action statement

By inserting the keywords in a sentence we are injecting a process into it by giving the action more agency. In Austin's [1] speech act theory terms, we would be transforming keywords into clear directive sentences, sentences that will have illocutionary power (show the meaning conveyed). A directive sentence clearly has injunctive power, signifies an order, an action to be executed through the illocutionary act and with the perlocutionary act (the actual effect) as a result. It is our hypothesis that keywords on the other hand, simply evoke the locutionary act (the literal words and their meanings), because the syntactico-semantic context is missing. While the coded language with its literal minimal units of meaning, color, and elliptical format has been thus far functional for, to quote Austin, "securing the uptake" and fulfilling what is being asked, we argue that the illocutionary and the perlocutionary acts are explicitly missing, and by using formats that convey them we will be making comprehension faster and more optimal in various contexts.

For the purposes of this first experiment, we will not start by testing action injunctive statements but information statements such as "Galleys extraction available in Flight" or "Expect high cabin rate". Information statements inform pilots of the availability of a function or a new situation arising (*cf.* Example 2). The information statement does not ask the pilot to perform anything. As such, it is not injunctive like the action statements, and does not have the elliptical dots format *(cf.* Example 1*)*. We will not be testing action statements in this first experiment as it is more difficult to test the real comprehension of a directive (with the absence of the previously mentioned perlocutionary function), unless the participants perform what is asked of them. Therefore, before attending to action statements, we will start by testing the hypothesis on constative information statements in initial coded language vs. a more natural form concerning reaction time and accuracy of comprehension.

Example 2. Example of an information statement

5.2   Participants

These sequences are tested on naïve participants, both native and non-native speakers of English in order to attest to their usability and confirm or deny our hypothesis on human comprehension in general, before we proceed to testing them on the real end users (pilots) with the real corpus. The participants are pre-tested for text comprehension of English in order to verify ad hoc the effect of English proficiency on comprehension and reaction times as a function of the two tested formats.

5.3   Procedure

In this first experiment we will not be using the exact aeronautical corpus terms as it will be difficult for naïve participants to understand technical jargon. Therefore, the stimuli will be made up of everyday life images and sentences such as *"The window is open"* that emulate the wording and intentions of our original corpus statements.

Participants will be presented with an image that describes the information statement that we want to test, e.g.: an image of an open window. The image will disappear and the information statement will appear. The participants have to press "yes" if the image is congruent with the statement (*The window is open* (natural) or *Window open* (coded)) or "no" if the image is incongruent (e.g.: *a closed door*) with the proposed format.

6   Conclusion

This paper's main issue is to investigate the basis of comprehension-oriented CNL evaluations, and moreover to question the scarcity of empirical evidence in the domain. The few relatively recent studies that have been conducted to substantiate the use of certain CNLs have not managed to significantly conclude that those CNLs have led to globally enhanced comprehension. We propose here a preliminary experimental protocol that we hope will provide us with some psycholinguistic insight on comprehension-oriented CNL evaluations, and perhaps empirically prove that natural language structure enhances comprehension in certain contexts. It would be interesting for the paper to be taken as a trigger for further discussion on evaluation principles.

References

1. Austin, J.L.: How to do Things with Words. Oxford University Press, Cambridge (1975)
2. Bisseret, A.: Psychology for man computer cooperation in knowledge processing. In: Masson, R.F.A. (ed.) IFIP 1983, Information Processing 1983 (1983)
3. Chervak, S.: The Effects of Simplified English on the Performance of a Maintenance Procedure. Master's Thesis. State University of New York (1996)
4. Chervak, S., Drury, C., Ouellette, J.: Simplified English for aircraft workcards. Proc. Hum. Factors Ergon. Soc. Annu. Meet. **40**(5), 303–307 (1996)
5. Condamines, A., Warnier, M.: Linguistic analysis of requirements of a space project and their conformity with the recommendations proposed by a controlled natural language. In: Davis, B., Kaljurand, K., Kuhn, T. (eds.) CNL 2014. LNCS, vol. 8625, pp. 33–43. Springer, Heidelberg (2014)
6. Crystal, D., Davy, D.: Investigating English Style. Longman, London (1969)
7. DuBay, W.: Principles of Readability. Impact Information, Costa Mesa (2004)
8. Eckert, D.: The Use of Simplified English to Improve Task Comprehension For non-native English Speaking aviation maintenance technician students. Doctoral Dissertation, West Virginia University, WV (1997)
9. Flesch, R.: How basic is basic English? Harper's Mag. **188**(1126), 339–343 (1944)
10. Fuchs, N.E., Schwitter, R.: Attempto controlled english (ace). arXiv preprint cmp-lg/9603003 (1996)
11. Fuchs, N.E., Schwitter, R.: Specifying logic programs in controlled natural language. In: Proceedings of CLNLP 1995, 16 pages, Edinburgh (1995)
12. Hart, S., Simpson, C.: Effects of linguistic redundancy on synthesized cockpit warning message comprehension and concurrent time estimation, pp. 309–321 (1976)
13. Hinson, D.E.: Simplified English—Is it really simple? In: Proceedings of the 38th International Technical Communication Conference (1988)
14. Kittredge, R.I.: Sublanguages and controlled languages. In: Mitkov, R. (ed.) The Oxford Handbook of Computational Linguistics, pp. 430–447 (2003)
15. Kiwan, D., Ahmed, A., Pollitt, A.: The effects of time-induced stress on making inferences in text comprehension. In: European Conference on Educational Research, Edinburgh, September 2000
16. Kuhn, T.: A survey and classification of controlled natural languages. Comput. Linguist. **40**(1), 121–170 (2014)
17. Lopez, S., Condamines, A., Josselin-Leray, A., O'Donoghue, M., Salmon, R.: Linguistic analysis of english phraseology and plain language in air-ground communication. J. Air Transp. Stud. **4**(1), 44–60 (2013)
18. McNamara, D., Louwerse, M., McCarthy, P., Graesser, C.: Coh-Metrix: capturing linguistic features of cohesion. Discourse Process. **47**(4), 292–330 (2010)
19. Shubert, K., Spyridakis, J.H., Heat, S.: The comprehensibility of simplified English in procedures. J. Tech. Writ. Commun. **25**(4), 347–369 (1995)

20. Simpson, C.A.: Effects of linguistic redundancy on pilot's comprehension of synthesized speech, pp. 294–308 (1976)
21. Spaggiari, L., Beaujard, F., Cannesson, E.: A controlled language at airbus. In: Proceedings of EAMT-CLAW03, pp. 151–159 (2003)
22. Stewart, K.: Effect of AECMA Simplified English on the Comprehension of Aircraft Maintenance Procedures by Non-native English Speakers. University of British Columbia (1998)
23. Temnikova, I.: Text Complexity and Text Simplification in the Crisis Management Domain. Ph.D. thesis, University of Wolverhampton (2012)
24. Van Oosten, P., Tanghe, D., Hoste, V.: Towards an improved methodology for automated readability prediction. In: 7th Conference on International Language Resources and Evaluation (LREC 2010). European Language Resources Association (ELRA) (2010)

Using CNL for Knowledge Elicitation and Exchange Across Story Generation Systems

Eugenio Concepción[1], Pablo Gervás[2], Gonzalo Méndez[2(✉)], and Carlos León[1]

[1] Facultad de Informática, Universidad Complutense de Madrid, Madrid, Spain
{econcepc,cleon}@ucm.es
[2] Instituto de Tecnología del Conocimiento,
Universidad Complutense de Madrid, Madrid, Spain
{pgervas,gmendez}@ucm.es

Abstract. Story generation is a long standing goal of Artificial Intelligence. At first glance, there is a noticeable lack of homogeneity in the way in which existing story generation systems represent their knowledge, but there is a common need: their basic knowledge must be expressed unambiguously to avoid inconsistencies. A suitable solution could be the use of a controlled natural language (CNL), acting both as an intermediate step between human expertise and system knowledge and as a generic format in which to express knowledge for one system in a way that can be easily mined to obtain knowledge for another system – which might use a different formal language. This paper analyses the suitability of using CNLs for representing knowledge for story generation systems.

Keywords: CNL · Story generation · Knowledge representation

1 Introduction

Natural languages allow human communication and knowledge transmission, and they provide an unbeatable expressiveness for concept modelling and structuring. However, for the same reasons, they are substantially complex for automatic processing.

Controlled Natural Languages (CNLs) can be considered as a balance between the expressiveness of the natural languages and the need for a formal representation that can be handled by a computer. A CNL is an engineered subset of natural languages whose grammar and vocabulary have been restricted in a systematic way in order to reduce both ambiguity and complexity of full natural languages [1].

Against this background, CNLs are attractive because of two reasons: first of all, since they are subsets of natural languages, they are naturally easier to write and understand by humans than formal languages; secondly, they can be translated automatically (and often deterministically) into a formal target language and then be used for automated reasoning [1]. CNLs offer an additional advantage: unlike formal languages that require some degree of consensus concerning

© Springer International Publishing Switzerland 2016
B. Davis et al. (Eds.): CNL 2016, LNAI 9767, pp. 81–91, 2016.
DOI: 10.1007/978-3-319-41498-0_8

their syntax, a CNL should be more suitable for different teams to understand each other and therefore to more easily conclude an agreement. Of course, the application of formal languages such as XML, JSON or RDF formats is easy to achieve when considering interchange formats between systems that have established an agreed data model. This is not normally the case between systems that address storytelling from different perspectives, which are likely to have radically different data models. Interchange formats such as XML, JSON, or RDF will require a different translation procedure to convert knowledge built based on one data model to knowledge built on a different data model. Converting a given data model to and from a common CNL would allow every system to make their knowledge available to all other systems for which such a translation procedure is available.

Story generation systems are a form of expression for computational creativity. Using the words of Gervás [2], a story generator algorithm (SGA) refers to a computational procedure resulting in an artifact that can be considered a story. The term story generation system can be considered as a synonym of storytelling systems, that is, a computational system designed to tell stories.

2   Related Work

Storytelling systems require the representation and manipulation of large amounts of knowledge. This involves not only the product itself – stories represented at various levels of detail – but also the knowledge resources that are required to inform the construction processes. This section explores some of the aspects that need to be represented and some examples of how controlled natural language might be applied in specific cases.

The context in which the proposal presented in this paper occurs involves: the complexity of knowledge representation required for story generation systems, the already proven suitability of CNL in storytelling systems, the difficulties in eliciting the required knowledge, and existence of storytelling systems that already comtemplate automated transformations across different representation formats as an integral part of their functionality.

2.1   Knowledge Representation in Storytelling Systems

There are multiple dimensions when considering knowledge representation for story generation. Gervás and León [3] provided a list of the most relevant classifications, and proposed their own list of suitable dimensions obtained from the different aspects of a narrative: discourse, simulation, causality, character intention, theme, emotion, authorial intention, and narrative structure.

From a historical perspective, formal languages have been the most common way of knowledge representation. The reason for using formal languages is simplicity: they have a well-defined syntax, an unambiguous semantics and are very convenient for automated reasoning. Particularly, in the field of automatic story generation, there is an abundance of examples of this kind.

TALE-SPIN [4] is one of the earlier story generators that produced stories about the inhabitants of a forest. It was a planning solver system that took as inputs a collection of characters with their corresponding objectives, found a solution for characters goals, and finally wrote up a story narrating the steps performed for achieving those goals. TALE-SPIN knowledge representation relied on Conceptual Dependency Theory [5]. TALE-SPIN output can be defined as a trace through a problem-solving process where the problems were limited to a specific area of knowledge, named the problem domain, which was defined by a set of primitives, a set of goal states or problems, and procedures for achieving these goals. All this knowledge was expressed as a formal language.

Minstrel [6] was a story generation system that told stories about King Arthur and his Knights of the Round Table. Its building units were a collection of goals and the plans to satisfy them. Story construction in Minstrel operated as a two-stage process involving a planning stage and a problem-solving stage which reused knowledge from previous stories. The knowledge representation in Minstrel used an extension of a Lisp library called Rhapsody, a tools package that provided the user with ways to declare and manipulate simple frame-style representations, and a number of tools for building programs that used them. Minstrel used Rhapsody for defining frames, schemas with slots and facets which represent story themes or morals, dramatic effects, world states, characters beliefs and affects.

Mexica [7] was a computer model designed to generate short stories about the early inhabitants of Mexico. It used several knowledge structures for supporting its storytelling model: an actions library, a collection of stories for inspiring the new ones, and a group of characters and locations. The story generation process took as input a file of primitive actions for creating an in-memory data structure after processing. It also created additional structures by transforming the file of Previous Stories into the Concrete, Abstract and Tensional Representations. The data structure built by the initial step was called Primitive Actions Structure, and it served as a repository for the primitive actions, which consists of an action name and several sets representing characters and their circumstances. Relations in Mexica representing emotional links and tensions between characters were modelled by means of formal languages in terms of three attributes: type (love or friendship), valence (positive or negative) and intensity. Mexica knowledge base also contained stories created by humans representing well-formed narratives, expressed as action sequences.

MAKEBELIEVE [8] was a short fictional story generation system that used a subset of common sense from the ontology of the Open Mind Common Sense Knowledge Base [9] for describing causality. Binary causal relations were extracted from these sentences and stored as crude trans-frames. MAKEBELIEVEs original knowledge base has been continued subsequently by the Open Mind Common Sense ConceptNet [10]. A trans-frame [11] is a type of diagram used for representing the common information related to an action. Minsky used the Trans primitives from Conceptual Dependency Theory [5] as inspiration for trans-frame concept. Hence, these data structures can be used for representing a stereotyped situation.

2.2   Use of CNL in Storytelling Systems

There is not a long record of uses of CNL in the context of storytelling.

Inform [12] was a toolset for creating interactive fiction. From version 7 on, Inform provided a domain-specific language for defining the primary aspects of an interactive fiction like the world setting, the character features, and the story flow. The provided domain-specific language used a CNL, similar to Attempto Controlled English [13].

The StoryBricks [14] framework was an interactive story design system. It provided a graphical editing language based on Scratch [15] that allowed users to edit both the characters features and the logic that drove their behaviour in the game. By means of special components named story bricks, users could define the world in which characters live, define their emotions, and supply them with items. Story bricks were blocks containing words to create sentences in natural language when placed together. They served to define rules that apply under certain conditions during the development of the story in the game.

In the extended ATTAC-L version [16], authors introduced a model which combined the use of a graphical Domain Specific Modeling Language (DSML) for modelling serious games narrative, ATTAC-L [17], with a CNL to open the use of the DSML to a broader range of users, for which they selected Attempto Controlled English [13]. It allows describing things in logical terms, predicates, formulas, and quantification statements. All its sentences are built by means of two word classes: function words (determiners, quantifiers, negation words, etc.) and content words (nouns, verbs, adverbs and prepositions). The main advantage is that Attempto Controlled English defines a strict and finite set of unambiguous constructions and interpretation rules.

3   Knowledge Elicitation for Storytelling Systems

Storytelling systems are extremely knowledge hungry. Generated stories are only as good as the knowledge they have been derived from. Given the thirst for knowledge of story generation systems, knowledge elicitation has always been a significant concern for researchers in this area.

Recent attempts have been made to address this problem via crowdsourcing [18]. In this work, a number of human authored narratives are mined to construct a *plot graph*, which models the author-intended logical flow of events in the virtual world as a set of precedence constraints between plot events. Typical narratives in natural language on a given topic collected via Amazon Mechanical Turk (AMT). The crowd workers are required to follow a simplified grammar and a number of restrictions that resemble closely a CNL. These narratives are parsed and merged into a combined representation in terms of plot graph for the domain being explored, which is later used to inform the process of constructing narratives.

In this regard, the Genesis system [19] can also be considered as an interesting example of knowledge mining from stories written in simplified English. Genesis was developed for studying the story understanding process, including the human ideological bias. It takes as inputs the text in Genesis English and a set of constraints representing the cultural and ideological context of the human reader that it will emulate. As a result, the system builds a graph-based representation using common sense rules. This knowledge structure allows the system not only to analyse problems, but also to answer questions and generate conclusions.

To inform the development of the Dramatis system for modelling suspense [20], O'Neill carried out an effort of knowledge engineering driven by methods adapted from qualitative research. The goal was to collect typical reader genre knowledge while simultaneously limiting engineer bias. The process was to acquire a corpus of natural language text and the conversion of that corpus into the knowledge structures required by Dramatis.

Although no CNL were strictly used for these tasks, the potential for their use in this context is clear.

3.1   Transformation Across Representations in Storytelling Systems

Gervás [21] attempts to model the procedures for composing narrative discourse from non-linear conceptual sources. It establishes algorithmic procedures for constructing a discourse – characterised by being a linear sequence of statements – to describe a set of facts known to have happened – which may involve events affecting a number of characters at different locations on overlapping periods of time. The composition procedure starts from a description of the set of events to be considered, produces an intermediate representation that captures the typical human view of events – restricted to what might have been perceived by a given character at a given moment in time – and proposes an algorithmic procedure to build a sequence of spans of narrative discourse, each capturing perception by an individual character. These spans of narrative discourse are rendered first as a simple conceptual description and then as pseudo-text. Table 1 shows examples of the original input as algebraic notation for a chess game, the conceptual representation of the composed discourse, and the pseudo-text rendering.

This conceptual description is built in such a way that the system can interpret it to reconstruct a version of the exhaustive description of the world that it started from. This is used by the system to validate the decisions taken during the composition of the discourse. The ability to convert automatically from one to another across different formats of knowledge representation can play a crucial role in the proposal described in this paper.

4   Using CNL for Knowledge Elicitation and Exchange Across Story Generation Systems

There are two important conclusions that can be extracted from the material presented so far.

First, that story generations are faced with a significant challenge of acquiring knowledge resources in the particular representation formats that they use. The difficulty of expressing human knowledge in formal languages is a considerable

Table 1. Original input as algebraic notation for a chess game, conceptual representation of the composed discourse, and pseudo-text rendering.

			The fifth black pawn was three squares north of the centre of the board. The fifth black pawn found the
		focalizes_on bp5	black left bishop was behind.
1. e4 e5	9. Nc3 Qe8	is_at lbb behind	The fifth black pawn found
2. Nf3 Nc6	10. Bxf6 gxf6	is_at bp4 same	the fourth black pawn was
3. Bc4 Bc5	11. Nh4 Kh8	is_at bp4 same	same. The fifth black pawn
4. O-O Nf6	12. Qh5 Be6	is_at bq behind	found the black queen was
5. h3 d6	13. Qxh6+ Kg8	is_at bk behind	behind. The fifth black
6. d3 O-O	14. Nd5 Bxf2+	move_to bp4 1/south	pawn found the black king
7. Bg5 h6	15. Rxf2 Qd7	4 later	was behind. The fifth black
8. Bh4 b6	16. Nxf6# 1-0	arrives_at rbk left	pawn saw the fourth black
		. . .	pawn moving south a square. Four days later, the fifth black pawn saw the black right knight arriving left.
			. . .

obstacle. In this context, the use of a CNL would provide the means for quicker development of required resources in a format easier to write for human experts. There are ongoing efforts to build such information via crowdsourcing and/or to learn this information via information extraction techniques, but all efforts along these lines either:

– have met with limited success,
– need to rely on huge amounts of hand annotation,
– require procedures of controlled edition very similar to CNL.

Second, that every story generation system defines its own format for knowledge representation, optimised to support its storytelling process. Although the development of a common formalism for representing knowledge would provide a major breakthrough, this is unlikely to happen in the near future (see Gervás and León [3] for a detailed discusion of the problems involved). Under the circumstances, the use of a CNL for codifying resources for storytelling systems might provide some relief. If authors of storytelling systems were to develop the initial version of their resources in a commonly agreed CNL, and then develop the appropriate automated transformations to generate knowledge in their own preferred format, the same resources written in CNL might be of use to researchers developing different storytelling systems. By simply writing the appropriate transformations into their own preferred format, much of the already available knowledge could be reused.

The main advantage of using a CNL is that it can be expressed by domain experts and then it can be translated to the variety of formal languages used in different systems. This feature allows the creation of a common language not only for expressing the different aspects involved in narrative generation, but also for exchanging knowledge resources across different storytelling systems. This might also pave the way for the development of common benchmarks for testing storytelling systems. A relevant conclusion mentioned by Gervás and León [3] is that the same information may be represented through different data structures without affecting its essence, or a data structure can be extended for representing additional types of information.

5   A Case Study: The STellA System

STellA (Story Telling Algorithm) [22,23] is a story generation system that controls and chooses states in a non-deterministically generated space of partial stories until it finds a satisfactory simulation of events that is rendered as a story. This simulation has been modeled as a knowledge intensive approach in which the whole world domain is explicitly represented as a simplistic view of a realistic environment. At each step, candidate updated versions of the current state are computed and the most likely ones are identified by computing their likelihood in terms of their plausibility and their narrative properties. Candidate partial stories are evaluated based on how well they satisfy a given set of constraints and how their tension curves compare with a set of target curves. The results of this process are used to decide when a partial story is promising and whether a story is finished.

5.1   Knowledge Representation in STellA

STellA is a resource hungry system. The underlying knowledge driving the generation has a big impact on the final quality of the output. One of the main characteristics of STellA is its heavy dependency on a core set of rules defining the whole universe the system is capable of producing stories about. The application of a CNL can highly improve the situation by letting external sources be added as knowledge.

The rules are considered part of the domain, and these basically operate by producing sequences of *snapshots* and *actions*. Snapshots are states of the world. A snapshot describes exact character positions, affinities, items, and every other detail of the world. Actions contain information about what led from the previous state to the current one. Then, a story is formally defined as show in Eq. 1.

$$story = \{(s_1, [a_{1,1}, a_{1,2}, \ldots, a_{1,n}]), \ldots, (s_z, [a_{z,1}, a_{z,2}, \ldots, a_{z,n})\} \qquad (1)$$

where s_x is a snapshot and $a_{i,j}$ is an action. Each pair is called a *state*, and a sequence of states form a story. Actions have their own vocabulary and correspond to specific structures like *take(character, item)* or *approach(character, place)*. Snapshots are defined according to a fixed ontology that structures the world. In STellA, the world is a matrix, and every entity fills exactly one cell. Big entities, as houses, are composed of small entities (bricks).

Each entity has its own set of attributes. Characters, in particular, are the most developed and detailed entities and are described in terms of properties commonly influencing narratives:

– physical properties for moving and interacting with the environment,
– affinities between characters,
– an internal representation of the world, which does not have to be the same as the real world. Characters use this for planning.
– Roles (moral tendency) and traits (special capabilities)

Rules, then, receive a current state (a pair of a snapshot and the actions that led to it from the previous one) and non-deterministically output a new set of actions. The non-deterministic aspect is not relevant for this discussion, and more information can be found on the literature [23]. The creation of rules is, then, driven by the data model.

5.2   Fundamental Features of a CNL for STellA

In order to design a CNL for the knowledge base of STellA, the data model must be covered by the language. While it is not a trivial task, the fact that the data model is well established and structured means the vocabulary can be easily described and expressed in natural language. In particular, the sentences to use must be able to describe preconditions and actions. Let us examine the following case:

When a character is alive and it moves north, its y coordinate has to be increased.

State information and actions are used in the sentence:

– *a character is alive* refers to a state. The rule is valid for *any* character that is alive.
– *it moves north* is an action, it corresponds to *move(?x, north)*.
– *its y coordinate must be increased* applies a change in the state: $y = y + 1$.

In order to transform the sentence into a rule, a template must be filled in like this:

precondition $\forall\ ?c\ \in\ Characters,\ alive(?c)$
action *move(?c, north)*
postcondition $?c.y\ \leftarrow\ ?c.y + 1$

Quantified state and action information must be addressed, which is, given the data model, doable. The problem is the generality needed in the kind of changes that must be applied in the postconditions. In this case, the simple *move* rule has updated the value of the y location component of the character *?c*. For more complex changes (like, for instance, traversing all available items in *?c*'s inventory), the CNL should cover a non-trivial set of constructions.

It is hypothesized that these complex constructions are, in general, mostly covered by a number of basic operations (traversing lists, accessing elements with given properties or finding elements in the world). Making this constructions atomic and suitable for composition can led to a simpler, reasonable expressive CNL. More in-depth study must be conducted in order to gain better insight on what this set of properties is.

6 Discussion

The construction of a CNL with the desired properties is a significant challenge. Namely because, to cover all the various aspects of representation relevant for storytelling systems as a whole, it would have to address at least all aspects described by Gervás and León [3]. From a simplified point of view, two major layers of representation can be considered. Firstly, an orchestration layer, whose main concern is related to the dynamic flow of the story. Secondly, a characterization layer, which is focused on representing the static features of the elements that define the story. The orchestration layer is related to the discourse sequence aspect, the causal aspect, and the intentional aspect. On the other side, the characterization layer includes the remaining aspects: the simulation aspect, the theme aspect, the emotional aspect, the authorial aspect, and the narrative structure aspect. It is both necessary and important to emphasise that these layers, and the aspects associated to them, are mutually interwoven. This means that changes in the data related to one aspect typically will cause changes in other aspects. For example, a change in the feelings of a character (emotional aspect) could determine his/her course of action, or modify his/her objectives (intentional aspect).

In addition, the use of a domain-specific glossary would serve not only for establishing a proper definition of the knowledge domain, but also for reducing the risk of polysemy. One of the potential issues with CNL is that they are not specifically designed to address word sense disambiguation. The definition of a CNL usually focuses on analysing just some key words that are relevant for building the discourse representation structure.

In a first-approach, it is possible to define a basic modelling for just some aspects of the narration. One suitable candidate could be the sequential aspect through the use of a planning modelling language, since a good part of the story generation systems architecture is based on a planner. A common language for modelling planners is PDDL (Planning Domain Description Language) [24], which is designed to formalize dynamic models, where actions guide the model through a series of state. This first step would consist of developing a match between the knowledge modelled in PDDL and a collection of primitives for describing the same information.

7 Conclusions and Future Work

This paper discussed the suitability of using a CNL for eliciting and exchanging knowledge in the context of a range of story generation systems. As shown above, there have been precedents of the use of CNL in the interactive storytelling domain with satisfactory results. The advantages of using a CNL for elicitation of knowledge resources had been demonstrated in the past. The potential for providing a compatible format for the exchange of knowledge across systems would be a major positive contribution to the field.

Future work involves not only the complete development of a CNL for covering the knowledge needed in this domain, but also the development of evaluation techniques that validate the suitability and portability of this representation over a wide range of story generation systems. In particular, a short term goal of the authors is to establish a CNL that can serve to develop knowledge resources that might as common ground data for a shared evaluation task for storytelling systems.

Acknowledgements. This paper has been partially supported by the project WHIM (611560) funded by the European Commission, Framework Program 7, the ICT theme, and the Future Emerging Technologies FET program.

References

1. Schwitter, R.: Controlled natural languages for knowledge representation. In: Proceedings of the 23rd International Conference on Computational Linguistics: Posters. COLING 2010, pp. 1113–1121. Association for Computational Linguistics, Stroudsburg (2010)
2. Gervás, P.: Story generator algorithms. In: The Living Handbook of Narratology. Hamburg University Press (2012)
3. Gervás, P., León, C.: The need for multi-aspectual representation of narratives in modelling their creative process. In: 5th Workshop on Computational Models of Narrative. OASIcs-OpenAccess Series in Informatics (2014)
4. Meehan, J.R.: Tale-spin, an interactive program that writes stories. In: Proceedings of the Fifth International Joint Conference on Artificial Intelligence, pp. 91–98 (1977)
5. Schank, R.C., Abelson, R.P.: Scripts, plans, and knowledge. In: Proceedings of the 4th International Joint Conference on Artificial Intelligence, vol. 1, IJCAI 1975, pp. 151–157. Morgan Kaufmann Publishers Inc., San Francisco (1975)
6. Turner, S.R.: Minstrel: A Computer Model of Creativity and Storytelling. Ph.D. thesis, University of California at Los Angeles, Los Angeles, CA, USA (1993). UMI Order no. GAX93-19933
7. Perez y Perez, R.: MEXICA: A Computer Model of Creativity in Writing. Ph.D. thesis, The University of Sussex (1999)
8. Liu, H., Singh, P.: Makebelieve: using commonsense knowledge to generate stories. In: Dechter, R., Sutton, R.S. (eds.) AAAI/IAAI, pp. 957–958. AAAI Press / The MIT Press (2002)
9. Singh, P., Lin, T., Mueller, E.T., Lim, G., Perkins, T., Zhu, W.L.: Open mind common sense: knowledge acquisition from the general public. In: Meersman, R., Tari, Z. (eds.) CoopIS 2002, DOA 2002, and ODBASE 2002. LNCS, vol. 2519, pp. 1223–1237. Springer, Heidelberg (2002)
10. Liu, H., Singh, P.: ConceptNet a practical commonsense reasoning tool-kit. BT Technol. J. **22**(4), 211–226 (2004)

11. Minsky, M.: Society of mind. Simon and Schuster, New York (1988)
12. Reed, A.: Creating Interactive Fiction with Inform. Cengage Learning, Boston (2010)
13. Fuchs, N.E., Schwertel, U., Schwitter, R.: Attempto controlled english – not just another logic specification language. In: Flener, P. (ed.) LOPSTR 1998. LNCS, vol. 1559, pp. 1–20. Springer, Heidelberg (1999)
14. Campbell, M.: A new way to play: make your own games. New Sci. **211**(2829) (2011)
15. Resnick, M., Maloney, J., Monroy-Hernández, A., Rusk, N., Eastmond, E., Brennan, K., Millner, A., Rosenbaum, E., Silver, J., Silverman, B., Kafai, Y.: Scratch: programming for all. Commun. ACM **52**(11), 60–67 (2009)
16. Van Broeckhoven, F., Vlieghe, J., De Troyer, O.: Using a controlled natural language for specifying the narratives of serious games. In: Schoenau-Fog, H., et al. (eds.) ICIDS 2015. LNCS, vol. 9445, pp. 142–153. Springer, Heidelberg (2015). doi:10.1007/978-3-319-27036-4_13
17. Broeckhoven, F.V., Troyer, O.D.: Attac-l: A modeling language for educational virtual scenarios in the context of preventing cyber bullying. In: 2nd International Conference on Serious Games and Applications for Health, pp. 1–8. IEEE, May 2013
18. Li, B., Lee-Urban, S., Johnston, G., Riedl, M.O.: Story generation with crowd-sourced plot graphs. In: Proceedings of the 27th AAAI Conference on Artificial Intelligence, AAAI 2013 (2013)
19. Winston, P.H.: The genesis story understanding and story telling system a 21st century step toward artificial intelligence. Technical report, Center for Brains, Minds and Machines (CBMM) (2016)
20. O'Neill, B.: A Computational Model of Suspense for the Augmentation of Intelligent Story Generation. Ph.D. thesis, Georgia Institute of Technology, Atlanta, Georgia (2013)
21. Gervás, P.: Composing narrative discourse for stories of many characters: a case study over a chess game. Literary Linguist. Comput. **29**(4), 511–531 (2014)
22. León, C., Gervás, P.: A top-down design methodology based on causality and chronology for developing assisted story generation systems. In: 8th ACM Conference on Creativity and Cognition, Atlanta, November 2012 (2011)
23. León, C., Gervás, P.: Creativity in story generation from the ground up: non-deterministic Simulation driven by Narrative. In: 5th International Conference on Computational Creativity, ICCC 2014, Ljubljana, Slovenia (2014)
24. McDermott, D., Ghallab, M., Howe, A., Knoblock, C., Ram, A., Veloso, M., Wilkins, D.: PDDL - the planning domain definition language (1998)

Towards a High-Level Controlled Language for Legal Sources on the Semantic Web

Adam Wyner[1](✉), Adeline Nazarenko[2], and François Lévy[2]

[1] University of Aberdeen, Aberdeen, UK
`azwyner@abdn.ac.uk`
[2] LIPN, Paris 13 University – Sorbonne Paris Cité & CNRS, Paris, France
`{adeline.nazarenko,francois.levy}@lipn.univ-paris13.fr`

Abstract. Legislation and regulations are required to be structured and augmented in order to make them serviceable on the Internet. However, it is known that it is complex to accurately parse and semantically represent such texts. Controlled languages have been one approach to adjusting to the complexities, where the source text is rewritten in some systematic form. Such an approach is not only costly, but potentially introduces alternative translations which may be undesirable. To navigate between the requirements and complexities, we propose and exemplify a high-level controlled language that serves as an XML representation for key components of legal content. The language tightly correlates to the source text and also facilitates analysis.

Keywords: Natural language simplification · Semantic annotation · Legal rules · Controlled languages · Semantic web

1 Introduction

The increasing complexity and integration of legal documents and regulations calls for rich legal content management. However, the complexity of legal language and regulations has long been understood to be an obstacle to the development of legal content management tools; for example, as discussed in [15], the complexity and ambiguity of the resulting parses and semantic representations make them difficult to evaluate for correctness as well as to exploit for experts in formal languages, *a fortiori* for legal analysts. The legal semantic web aims at giving a uniform access to legal sources, whatever form they may take or the institution that published them. This is traditionally supported by the definition of a metadata vocabulary and the semantic annotation of the sources. Beyond documents and topic-based annotations, however, legal experts must have direct access to the rules contained in documents and their supported interpretations. This calls for a rich and structured annotation of the rule text fragments.

However, problematics arise from the tensions between the complexities of legal natural language, the requirements of legal professionals, and the specifications of formal or machine-readable languages. In this paper, we attempt to

© Springer International Publishing Switzerland 2016
B. Davis et al. (Eds.): CNL 2016, LNAI 9767, pp. 92–101, 2016.
DOI: 10.1007/978-3-319-41498-0_9

moderate the tensions using a simplified, yet useful controlled language (CL) to mark up the source text. The novel, significant contribution of this paper is the advocation for an analysis and annotation of legal sources using structured annotations, which is our high-level CL (hCL), on the source text. This hCL leaves the source text *in situ*. We claim that the annotations can be semi-automatically associated with NL expressions, and moreover, that the annotations can be associated with representations in XML. In the advocated approach, one rule in the source text can be annotated with different CL statements so as to represent different interpretations, thus leaving it up to the analyst to resolve ambiguities. Furthermore, our hCL focuses on the semantic structure of the rules, providing an abstract representation of the components of a proposition rather than a collection of annotations; as an analysis of sentence components, it is similar to a parser, yet it focuses on semantic annotations that are key to rules. The approach combines the source text for reference, the controlled language annotations for experts, and the semantic representation for Semantic Web querying.

To ground our discussion and provide a running example, we use a corpus that was previously reported in [16], which is a passage from the US Code of Federal Regulations, US Food and Drug Administration, Department of Health and Human Services regulation for blood banks on testing requirements for communicable disease agents in human blood, Title 21 part 610 Section 40. We present a running example from this corpus.

In the remainder of the paper, we outline existing research to contrast with our proposal (Sect. 2). We sketch our annotation approach to our hCL in Sect. 3. We present a specification of the hCL (Sect. 4), then present an incremental methodology by example for annotating legal sources with hCL statements (Sect. 5). The paper closes with a summary and some indications of future work.

2   Related Work

The complexity of legal language and regulations has long been understood to be an obstacle to the development of legal content management tools. Attempts have been made to parse and automatically formalize legal texts. For instance, C&C/Boxer [3] has been applied to fragments of regulations [15]. C&C/Boxer is a wide coverage parser that feeds a tool which generates semantic representations (essentially in First-order Logic). However, as discussed in [15], the complexity and ambiguity of the resulting parses and semantic representations make them difficult to evaluate for correctness as well as to exploit for experts in formal languages, *a fortiori* for legal analysts.

Controlling the legal sources has been proposed as an alternative approach. Efforts are made to clarify and simplify the legal language when drafting (*e.g.* in favor of "Plain English", to ease translation [12], or to avoid ambiguity and clumsiness [8]). More formally, a wide range of controlled languages (CL) has been proposed [9], with the idea that controlled statements would be easier to parse but still be meaningful and manageable. *Attempto Controlled English (ACE)* defines unambiguous readings of quantifier scopes and anaphora as well as prohibits ambiguous syntactic attachments, thus enabling a parse and translation into predicate logical formulae. The *Oracle Policy Modelling* (OPM) system [4] is designed to parse structured sets of controlled sentences and make rulebases available online. *Semantics of Business Vocabulary and Business Rules* (SBVR) has been specifically designed to model business rules [13]: it provides elements of a pattern language and a description of *SBVR-Structured English* to express rules in a form that can be checked by human experts. ACE, OPM, and SBVR try to systematize the NL to CL translation by proposing alternative formulations for unwanted constructions. However, when the source regulations get more complex, the NL to CL translation either fails or gives a formal result, with explicit scopes and qualifiers, which can be difficult to read and even harder to adjudicate for correctness.

A third approach relies on the semantic annotation of legal texts without diving into the detailed syntactic complexity of legal sentences. Annotations are made at the paragraph level, making use both of a high level legal ontology and a specific domain ontology [1]. Provision level annotations are given, which rely on a general model of relationships between normative provisions [6,14]. The provision collection is encoded in RDF-OWL and can be queried using SPARQL. In a similar vein, the LegalRuleML mark-up language is designed to represent legal rules for the semantic web [2].

These approaches share a common disposition – the source legal language must itself be normalized, transformed, and disambiguated in order to be systematically represented. This may not be feasible without unduly constraining the scope of analysis and of interpretation. A pragmatic proposal is to combine the controlled language and the semantic annotation approaches as initially proposed in [10], which provides some content of semantic annotations and fixes the interpretation of underlying fragments of legal sources. This approach builds on SemEx methodology, which was designed to annotate business regulations by business rules through an iterative rewriting process, ideally until a CL form is obtained [7]. However, a full specification of CL seems problematic. In this paper, we focus on key textual components to represent the main legal features rather than the details of domain terminology.

3 Annotating Legal Content with hCL

Formalization of legal documents yields representations that support content management (indexing and search, merge, comparison, update of documents) and legal reasoning (*Is it necessary to test X for Y?* and *If X, then Y*). However, completely formalising the content of legal documents is a distant goal, due to legal and domain-specific terminology, long and complex sentences as well as ambiguities. It would, nonetheless, be useful for legal content management and reasoning to provide a degree of formalization as structured annotations. We develop the formalization *pragmatically* and *partially*; it is pragmatic as we only annotate those components as needed for the analysis of rules, and it is partial in that we allow mixing of annotations and unannotated source text.

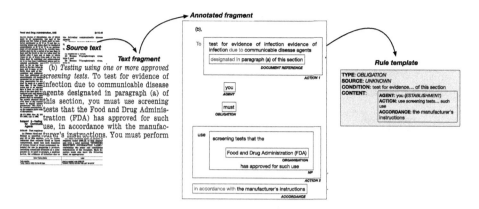

Fig. 1. Example of an annotated rule statement and of the semantic pattern that can be derived from the hCL annotation.

We focus our analysis on legal rules and their components. To illustrate our approach, we show an analysis of our running example below as in Fig. 1, which is further discussed in Sect. 4:

(b) To test for evidence of infection due to communicable disease agents designated in paragraph (a) of this section, you must use screening tests that the Food and Drug Administration (FDA) has approved for such use, in accordance with the manufacturer's instructions.

The analysis relies on the following intuitions:

- Formalization in hCL adds annotations to the source text, enriching it and leaving it unchanged.
- hCL aims at providing simplified and semantically more explicit versions of the components of rule statements and their integration as rules.
- Formal statements can be expressed through form-based semantic structures for rules. These forms are usually filled with high level annotations, either because there is a straightforward correspondence between the annotations and the semantic roles or because they correspond to a characteristic pattern of annotations. In any case, experts play a key role in the formal labeling of patterns.
- Annotated analyses can be *folded* and *unfolded* so that annotations can accommodate various granularities of formalization. A fully folded analysis is just the annotations (perhaps along with some keywords); a fully unfolded analysis is just the source text.

Interpretation of legal texts is important in legal reasoning, where there always remains room for interpretation. Annotating therefore amounts to specifying an interpretation through the selection of the most important fragments of the source regulation and the clarification of the semantic structure of the rules,

1. [X]AGENT [must]MODAL [use screening tests]ACTION [in accordance with the law L]ACCORDANCE.

2. [The law L]SOURCE [makes [X]AGENT an obligation]MODAL [to use screening tests]ACTION.

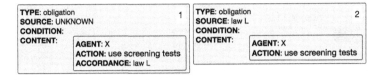

Fig. 2. Alternative annotation of an ambiguous source sentence: The ambiguity concerns the attachment of the prepositional phrase *in accordance with the law L* to the modal or main verb (Readings 1 and 2, resp.).

e.g. the relationships amongst the fragments. Figure 2 shows how the annotation in hCL highlights two alternative readings of an ambiguous text fragment.

Note that the original terms (*e.g. screening tests* in Fig. 2) are preserved in the annotations, so that their applicability to actual cases can be discussed in legal terms.

We propose that the content of the annotation be a linguistic expression to preserve readability. In the hCL design, focus is put on the constituent clauses of the rule statements, which are associated with an explicit and unambiguous semantics (see Fig. 2), leaving aside the detailed parsing of the constituents. These may remain unanalyzed (*e.g. use screening tests* in Fig. 2). This approach hides ambiguities and complexities of the lower level of analysis (*e.g. the anaphoric expression *this section*) to highlight the main structure of the rule statements. Our presumption is that such ambiguities and complexities are either left unanalyzed or are treated by auxiliary processing components. Yet, our approach remains flexible: two different analysts may propose compatible readings even if one is more coarse-grained than the other.

Annotations can be exploited for content management and legal content mining with respect to rules. High level annotations homogenize linguistic variation and are used for search, instead of searching by keywords. This allows for answering queries like: *Which rules appear in a document?*, *Do the analysts agree on the interpretation of a specific rule statement or more generally on a section of a document?*, *What are all the rules that concern X?*, and *What are the rules involving a given action?*.

4   Specification of hCL

This section presents hCL and describes how this language can be used to analyze rule statements so as to make explicit the overall semantic structures of the rules. The goal is not to have a complete analysis of the rule statements, but rather to structure and index the rules in a systematic and explicit way so as to enable users to mine the legal sources as semi-structured, semantically annotated documents, yet leaving the source text unchanged.

We assume here (and discuss in Sect. 5) that we can semi-automatically identify semantic annotations, *e.g.* AGENT, with relevant syntactic phrases, *e.g.* noun phrase. Thus, the basic terminology of hCL is: AGENT, THEME, DEONTIC, ACTION, STATE, ACCORDANCE.

Each of these may be realized by a variety of syntactic expressions, so abstracting over linguistic heterogeneity. The AGENT and THEME correlate to noun phrases, DEONTIC correlates to various expressions of deontic modality, *e.g. must, may, prohibited, obligated,* ACTION and STATE are verb phrases, and ACCORDANCE is an adjunct phrase. The ACTION and STATE annotations, as verb phrases, incorporate their verb phrase internal arguments (*i.e.* direct objects and indirect objects). Given the underspecified approach adopted here, modals, actions, and states can be positive or negative.

The simplest rule pattern of hCL is[1]:

$$\text{RULE} \leftarrow \text{AGENT DEONTIC ACTION} \tag{1}$$

In this schema, AGENT, DEONTIC and ACTION elements refer to text fragments that have been annotated as such. Over 30 rule statements, 16 occurrences of this rule schema can be found in our text example. For instance, in *You must use screening tests*, where *you* is the AGENT, *must* is DEONTIC, and *use screening tests* is the ACTION.

This is a simple example. Given complex legal statements, the correspondence is often more complex: in *You must use screening tests that the Food and Drug Administration (FDA) has approved for such use, in accordance with the manufacturer's instructions.*, *you* is the AGENT, *must* is the DEONTIC, *use screening tests that the Food and Drug Administration (FDA) has approved for such use* is the ACTION, and *in accordance with the manufacturer's instructions* is the ACCORDANCE. In the example text, simple rules on average 44 words long (between 22 and 73).

While we can homogenize some linguistic variation, there are other variations we want to explicitly represent such as the active-passive (semantically annotated as ACTION and STATE) and the optionality of adjuncts. We assume that the AGENT of a passivised verb phrase (here indicated as STATE) is not a verb phrase internal argument and is optional. The ACCORDANCE annotation is, as an adjunct phrase, optional. Thus, rule pattern 1 can be revised to:

$$\text{RULE} \leftarrow \text{AGENT DEONTIC ACTION [ACCORDANCE]} \tag{2}$$

$$\text{RULE} \leftarrow \text{THEME DEONTIC STATE [AGENT][ACCORDANCE]} \tag{3}$$

As discussed above, legal rule statements can be complex and of various forms. It is not our intention to incorporate into the hCL all possible constructions, variants, and terminologies. We keep our annotations to a small feasible set and allow key words in the text *in-situ* to co-occur with the annotations (recalling lowercase words, such as *if* and *provided that*, refer to key words in the source documents). Such a mixed approach still allows the annotations to be useful for content management and reasoning without

---

[1] We have the following typographic conventions: capitalized elements (AGENT) refer to high level annotated textual fragments, brackets indicate optional elements, and lowercase words (*if*) refer to actual key words to be found in the source documents.

fully spelling out all details of the language of the source text. We allow that annotations can be further specified, *e.g.* RULE$_{perm}$ is a rule annotation where the DEONTIC is a *permission*:

$$\text{RULE} \leftarrow if \text{ AGENT ACTION, RULE} \tag{4}$$

$$\text{RULE} \leftarrow if \text{ THEME STATE, RULE} \tag{5}$$

$$\text{RULE} \leftarrow \text{RULE}_{perm} \text{ } provided \text{ } that \text{ AGENT ACTION} \tag{6}$$

$$\text{RULE} \leftarrow \text{RULE}_{perm} \text{ } provided \text{ } that \text{ THEME STATE} \tag{7}$$

All together these patterns cover 4/5 of the rule statements of our source text. We give few examples below:

1. [[you, an establishment that collects blood or blood components]$_{AGENT}$, [must]$_{DEONTIC}$ [test each donation of human blood or blood component intended for use in preparing a product for evidence of infection]$_{ACTION}$]$_{RULE}$

2. [If [you]$_{AGENT}$ [ship autologous donations to another establishment that allows autologous donations to be used for allogeneic transfusion]$_{ACTION}$, [[you]$_{AGENT}$ [must]$_{DEONTIC}$ [assure that all autologous donations shipped to that establishment are tested under this section]$_{ACTION}$]$_{RULE}$]$_{RULE}$

3. [If [a filling]$_{THEME}$ [fails to meet the requirements of the first repeat test]$_{STATE}$, [[a second repeat test [may]$_{DEONTIC\_PERM}$ be conducted on the species which failed the test]$_{ACTION}$]]$_{RULE}$, provided that [50 percent of the total number of animals in that species]$_{THEME}$ [has survived the initial and first repeat tests]$_{STATE}$.

4. [you]$_{AGENT}$ [must]$_{DEONTIC}$ [use screening tests that the Food and Drug Administration (FDA) has approved for such use]$_{ACTION}$, [in accordance with the manufacturer's instructions]$_{ACCORDANCE}$

Of course, some sequences of categories might be ambiguous. For instance, in Example 4 and as discussed in Fig. 2, the accordance phrase could be associated either with the action specifically or with the deontic modal, that is to the rule itself. In the former case, rule 2 applies, but the latter one calls for an additional specification:

$$\text{RULE} \leftarrow \text{RULE ACCORDANCE} \tag{8}$$

hCL is designed to identify the overall semantic structures of the rule statements expressed in legal documents[2]. We make no claim that the few rules above are exhaustive and complete, covering all legal rule statements; but we do claim that, with respect to our corpus, most of the statements can be explained with a limited set of rules. This high level language leaves aside the actual parsing of the texts as well as deep, detailed semantic annotation, since long text fragments can be annotated as single hCL components. This illustrates the pragmatic approach that we have adopted to tackle legal language: even rough annotations are useful in a semantic web perspective. At the high level, a statement is described as a sequence of hCL categories and key words. Attachment ambiguities appear when two different rules apply on the same fragment (as for the ACCORDANCE phrase above).

In this section, we have described basic rule structures using a small set of annotations. The following section explains how the basic elements of the rules can be identified in order to annotate the components of the rule.

---

[2] These structures can then be transformed into attribute-value structures as the one presented in Fig. 1 but we do not develop that point here.

5   Annotation Support

The initial annotation of the legal texts requires attention from analysts, but a range of resources are available to support that task, which we outline in this section. These cover not only rule identification, but also the identification of relevant components of rules.

Annotation Guidelines. Annotation guidelines must be produced to explain to analysts how legal texts should be annotated. For a given text fragment, all analysts do not have to produce exactly the same annotations but they have to produce compatible annotations which differ only in the granularity of their descriptions. The guidelines must:

- list and define all the allowed semantic categories, with positive and negative examples to illustrate how the categories can or should be used in annotation;
- explain how to handle complex markup issues such as discontinued elements or annotation embeddings;
- specify how the quality of the annotation can be evaluated and assessed.

Terminological Analysis. Key domain terminology can be readily identified from existing terminological list (*e.g. blood transfusion*) or using existing terminological term extraction tools (*e.g.* TermRaider [11] or Termostat [5]). These terms, which are often called "open-textured terms" by legal analysts play an important role in legal interpretation. As legal reasoning consists in applying rules to facts, the classification of the actors, components of a situation, and so on plays a central role. Semantic tagging must therefore identify the key domain terms on which interpretation is based. The goal is not to propose a formal definition of those terms, as the subtleties of the different cases and the open-world assumption usually prevents the complete formalization of legal policies. The goal is simply to highlight them and make explicit their variations.

Interactive Annotation Tool. It is essential to provide analysts with a dedicated, interactive annotation tool. Many different generic tools can be exploited, *e.g.* the GATE, NLTK, or UIMA[3]. Key to using such tools is the application of the finite set of annotations as described in Sect. 4. Once a sequence of categories and keywords matches a right hand part of a rule, the analysts can tag the whole statement as a rule. If the rule is ambiguous, the analyst may be warned that he/she has to choose among alternatives.

Local Grammars. We have kept our discussion to high level rule components. Clearly, these are not sufficient to cover all of the textual phenomena that may be relevant to a fuller analysis of the text. For this, local grammars, which are grammars designed for subtasks of the overall analysis, may be designed to help analysts identifying the key elements of legal language, such as the markers and phrasal structures that introduce complements and adjuncts, such as *due to* and *except*, or legal terminology (*e.g.* references to textual elements). We may assume that local grammars are applied on

---

[3] https://gate.ac.uk/
http://www.nltk.org/book/
https://uima.apache.org/index.html.

documents that have already been POS-tagged and chunked so as to identify the borders of the elements that have been identified.

This pipeline of NLP processes support the analyst, but leave the task of semantic tagging under her control.

6 Summary and Future Work

The paper has advocated, motivated and exemplified a pragmatic approach to the analysis and annotation of complex legal texts. The approach combines the benefits of controlled languages – to give manageable although simplified descriptions of legal content – and of semantic annotation – to maintain a tight correlation with the source texts. It was pragmatically designed to help analysts publish legal sources and share interpretations on the semantic web.

For future work, we plan to apply the analysis to the larger regulation from which the sample is drawn, modifying it as required; for instance, the initial fragment could be extended to other constructions, *e.g.* exceptions and conditionals [16]. We would add tool support, *e.g.* contextually relevant pop-up annotation alternatives along with the option to create new, which would be essential to control for annotation variation. Evaluations with respect to inter-annotator agreement and users (e.g. querying) would help to establish the strengths or weaknesses of the approach and tools for the intended texts and user group.

References

1. Asooja, K., Bordea, G., Vulcu, G., O'Brien, L., Espinoza, A., Abi-Lahoud, E., Buitelaar, P., Butler, T.: Semantic annotation of finance regulatory text using multilabel classification. In: LeDA-SWAn (to appear, 2015)
2. Athan, T., Boley, H., Governatori, G., Palmirani, M., Paschke, A., Wyner, A.: OASIS LegalRuleML. In: ICAIL, pp. 3–12. Rome, Italy (2013)
3. Bos, J.: Wide-coverage semantic analysis with boxer. In: Proceedings of Semantics in Text Processing, Research in Computational Semantics, pp. 277–286. College Publications (2008)
4. Dayal, S., Johnson, P.: A web-based revolution in Australian public administration. J. Inf. Law Technol. 1, online (2000)
5. Drouin, P.: Term extraction using non-technical corpora as a point of leverage. Terminology **9**(1), 99–115 (2003)
6. Francesconi, E.: Semantic model for legal resources: annotation and reasoning over normative provisions. Semant. Web **7**(3), 255–265 (2014)
7. Guissé, A., Lévy, F., Nazarenko, A.: From regulatory texts to BRMS: how to guide the acquisition of business rules? In: Bikakis, A., Giurca, A. (eds.) RuleML 2012. LNCS, vol. 7438, pp. 77–91. Springer, Heidelberg (2012)
8. Höfler, S.: Legislative drafting guidelines: how different are they from controlled language rules for technical writing? In: Kuhn, T., Fuchs, N.E. (eds.) CNL 2012. LNCS, vol. 7427, pp. 138–151. Springer, Heidelberg (2012)
9. Kuhn, T.: A survey and classification of controlled natural languages. Comput. Linguist. **40**(1), 121–170 (2014)
10. Lévy, F., Nazarenko, A., Wyner, A.: Towards a high-level controlled language for legal sources on the semantic web. In: LeDA-SWAn (to appear, 2015)

11. Maynard, D., Li, Y., Peters, W.: NLP techniques for term extraction and ontology population. In: The Conference on Ontology Learning and Population, pp. 107–127. IOS Press, Amsterdam (2008)

12. Meunier, M., Charret-Del Bove, M., Damette, E. (eds.): La traduction juridique: points de vue didactiques et linguistiques, 333 pages. Publications du Centre d'Etudes Linguistiques (2013)

13. OMG: Semantics of business vocabulary and business rules. Formal specification, v1.0. Technical report, The Object Management Group (2008)

14. Peters, W., Wyner, A.: Extracting hohfeldian relations from text. In: JURIX, Frontiers in Artificial Intelligence and Applications, vol. 279, pp. 189–190. IOS Press (2015)

15. Wyner, A., Bos, J., Basile, V., Quaresma, P.: An empirical approach to the semantic representation of law. In: JURIX, pp. 177–180. IOS Press, Amsterdam (2012)

16. Wyner, A., Peters, W.: On rule extraction from regulations. In: JURIX, pp. 113–122. IOS Press (2011)

The Controlled Natural Language of Randall Munroe's *Thing Explainer*

Tobias Kuhn(⊠)

Department of Computer Science,
VU University Amsterdam, Amsterdam, Netherlands
kuhntobias@gmail.com

Abstract. It is rare that texts or entire books written in a Controlled Natural Language (CNL) become very popular, but exactly this has happened with a book that has been published last year. Randall Munroe's *Thing Explainer* uses only the 1'000 most often used words of the English language together with drawn pictures to explain complicated things such as nuclear reactors, jet engines, the solar system, and dishwashers. This restricted language is a very interesting new case for the CNL community. I describe here its place in the context of existing approaches on Controlled Natural Languages, and I provide a first analysis from a scientific perspective, covering the word production rules and word distributions.

1 Introduction

The recent book *Thing Explainer: Complicated Stuff in Simple Words* [12] by Randall Munroe (who is most well-known for his *xkcd* webcomics) is a very interesting case for the research field of Controlled Natural Languages (CNL) [11]. It is "a book of pictures and simple words [...] using only the ten hundred words in our language that people use the most" [12] and it is both, fun and totally serious. The quote is from the introduction of the book, and therefore it is itself written in this language of only the 1'000 most commonly used English words (and so is, of course, the title of the book). The following paragraph is another example from the section about nuclear power plants, explaining radioactivity:

> The special heat is made when tiny pieces of the metal break down. This lets out a lot of heat, far more than any normal fire could. But for many kinds of metal, it happens very slowly. A piece of metal as old as the Earth might be only half broken down by now [12].

The book has attracted substantial popular interest and press coverage, probably more so than any other book written in a Controlled Natural Language in the recent past, or possibly ever. It has received very positive reviews from prestigious sources such as The New York Times [3], The Guardian [1] ("At some points, this produces passages of such startling clarity that one forgets there was ever anything difficult to understand about these phenomena."), and

© Springer International Publishing Switzerland 2016
B. Davis et al. (Eds.): CNL 2016, LNAI 9767, pp. 102–110, 2016.
DOI: 10.1007/978-3-319-41498-0_10

Bill Gates [8] ("a brilliant concept"), in addition to an interview in New Scientist [10]. But arguably the most flattering review is the one that appeared in The Huffington Post [9], because the journalist himself used the book's controlled language to write the entire review in it! ("So I thought I'd try to tell you a little about this new book the same way, using just those few words.") The first edition consisted of 300'000 printed copies [3], 34'000 of which were sold in the first week alone,[1] and at the time of writing the book is in the top 20 of best selling books at Amazon in the category *Science & Math*.[2] This popularity alone makes it an interesting and important CNL to have a closer look at.

The language of *Thing Explainer* is also interesting because of its intriguingly simple restriction applied on top of the English language, namely to use only the top 1'000 most often used words. This kind of simplicity is only rivaled among existing CNLs by the language E-Prime [5], whose only restriction is that the verb *to be* is forbidden to use. The fact that the language's restricted vocabulary is not quite as simple as it looks at first, as we will discover below, does not make the concept less intriguing. Most reviewers and readers seem to agree that Randall Munroe succeeds in proving that virtually everything can be explained in an understandable fashion with this so heavily restricted vocabulary.

2   Language Analysis

Even though the book and its language have become very popular, not much has been written about the details of the language, the connection to other similar languages, and the precise rules that underlie it. Randall Munroe himself introduced the language in the book with only a few sentences. It is therefore worth taking a closer look here.

2.1   Similar Languages

The new language of *Thing Explainer* is similar to some of the earliest English-based CNLs. Basic English [13] was arguably the first such language, presented in 1930. It restricts the allowed words to just 850 root words, but many variations of the language exist. The chosen words and the rules for their application are much more structured than the *Thing Explainer* language, however, allowing for only 18 verbs and imposing substantial restrictions on the grammatical structures within which these words can be used. In this sense, Special English [14] — arguably the second oldest English-based CNL — is more similar. It defines no grammatical restrictions and does not restrict the number of verbs so drastically. It is based on a list of 1'500 words, and has been used since 1959 by the Voice of America, the official external broadcast institution of the United States. In both

---

[1] http://www.publishersweekly.com/pw/by-topic/industry-news/bookselling/article/ 68882-the-weekly-scorecard-tracking-unit-print-sales-for-week-ending-november- 29-2015.html, retrieved on 9 April 2016.

[2] http://www.amazon.com/Best-Sellers-Books-Science-Math/zgbs/books/75/, retrieved on 7 April 2016.

cases, the words on the list are carefully selected and not just chosen by their frequency in English texts, unlike the *Thing Explainer* language. As another difference, Basic English and Special English define the category for the words on the list, such as *noun* and *verb*, and allow the words only to be used in the given category. Other similar languages have a more technical background, such as ASD Simplified Technical English (ASD-STE) [4], which also restricts the allowed words and grammatical structures for the aerospace domain, and there are many other similar languages [11].

2.2 Language Properties

According to the PENS classification scheme, which I proposed in my survey on the topic [11], the language of *Thing Explainer* has the same type as Special English, which is $P^1E^5N^5S^1$. This is also the type of full unrestricted English, meaning that the restrictions of the *Thing Explainer* language do not make it considerably different according to the dimensions of the PENS scheme: precision, expressiveness, naturalness, and simplicity. It is not considerably more precise than full English, because no semantic restrictions come with the restricted vocabulary and the grammar is not restricted at all, and therefore the vagueness and ambiguity of natural language is not significantly mitigated. In terms of expressiveness and naturalness, on the other hand, the power of full English is retained: Randall Munroe comes at least close to proving that basically everything that can be expressed in full English can be expressed in the restricted language as well, in a way that is maybe sometimes funny but always fully natural. With respect to the last dimension, the *Thing Explainer* language is simple when full English can be taken for granted, but it is not significantly simpler than the full language when it has to be described from scratch. Apart from Special English, other CNLs in the same PENS class are E-Prime, Plain Language, and IBM's EasyEnglish [11].

2.3 Word List

The list of 1'000 words was assembled by manually merging the word frequency lists from several corpora. Randall Munroe reports it like this:

> I spent several months going back over a bunch of different lists and generating some of my own based on the Google Books corpus and even my own email inbox. Then I combined the lists and where they disagreed I just let my sense of consistency be the tie-breaker [10].

Conjugated forms of verbs and plural forms of nouns are not listed separately but merged with the plain word form, as Randall Munroe explains: "I count different word forms—like 'talk,' 'talking,' and 'talked'—as one word." This does not apply though to comparative and superlative forms of adjectives (*good*, *better*, and *best* are all separate entries), adverb forms of adjectives (*easy* and *easily* count as two words), or pronouns (*I*, *me*, *my*, and *mine* are separate entries too).

This results in a list of 998 words, which is part of the book, where the missing two words are explained by the fact that "there's a pair of four-letter words that are very common, but which I left off this page since some people don't like to see them."

2.4   Word Production Rules

Now that we know how the words ended up on the list, let us have a look at how they are supposed to be used from there to write texts such as the book we discuss here. As there are no grammar restrictions, this boils down to selecting word forms from the list and applying word production rules to arrive at related word forms. Randall Munroe's own description of these production rules is very short and not very precise (the first sentence has been quoted above already):

> In this set, I count different word forms—like "talk," "talking," and "talked"—as one word. I also allowed most "thing" forms of "doing" words, like "talker"—especially if, like "goer," it wasn't a real word but it sounded funny.

As this description leaves a lot of room for interpretation, we can use the corpus of word forms observed in *Thing Explainer* to reverse engineer the specific rules at work. Doing so, we arrive at no less than 13 rules, listed here roughly in decreasing order of how naturally they follow from the above description:

1. The word forms on the list of the 1'000 most often used words.
2. All conjugation forms of verbs on the list. This includes third singular present (*-s*), past (*-ed*), and infinitive from (*-ing*), including irregular forms.
3. Noun forms built from verbs on the list by *-er*, for example *carrier*.
4. The plural forms of nouns on the list (*-s*), for example *things*, including irregular forms like *teeth*. This rule can also be applied to the word *other* to produce *others*, even though it is not a noun.
5. Comparative (*-er*) and superlative forms (*-est*) built from adjectives on the list, for example *smaller* or *fastest*, and including irregular forms like *worse*.
6. Adjective forms built from nouns on the list by *-y* or *-ful*, for example *pointy* or *colorful*. (The word *colorful* is in fact the only one in *Thing Explainer* derived from this *-ful* production rule.)
7. Adverb forms built from adjectives on the list by *-ly*, for example *normally*.
8. Noun forms built from adjectives on the list by *-ness*, for example *thickness*.
9. Different case and possessive forms of pronouns on the list: *they* for *them*, *us/ours* from *we/our*, and *his* from *he*.
10. Verb forms of nouns on the list and noun forms of verbs on the list when the two forms are similar but not equal, such as *thought* from *think*, and *live* from *life*, including deduced forms like *thoughts* and *living*. (These two cases are in fact the only ones observed for this rule in *Thing Explainer*.)
11. More basic word forms for words on the list, such as nouns from which adjectives on the list were built (*person* from *personal*) and verbs from which nouns on the list were built (*build* from *building*). (Again, these two cases are the only two observed instances of this rule.)

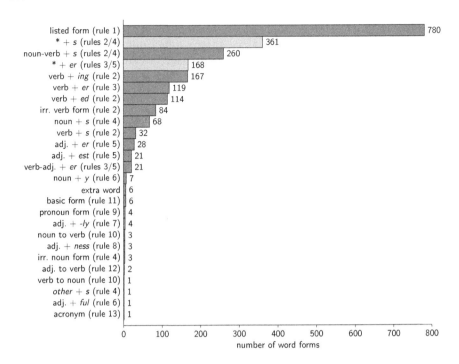

Fig. 1. The origin of the word forms found in *Thing Explainer* with respect to the different production rules. The total number of such word forms is 1736.

12. Verb forms built from adjectives on the list, such as *lower* from *low*, including conjugated forms like *lowering* and *lowered*. (This example is again the only observed instance of the rule.)
13. Common acronyms for words on the list, such as *TV* for *television*. (This example is also the only instance.)

It is not completely clear though, whether some of these later rules point to mistakes rather than features. The confusion of *TV* and *television*, for example, might just be a mistake and not a feature of the language.

Even with these rules, six words remain that are used in *Thing Explainer* but are not allowed according to these rules: *some*, *mad*, *hat*, *apart*, *rid*, and *worth*. It seems that the first one, *some*, should be on the top 1'000 list, but was accidentally omitted. The omission of *they* from the list seems to be a similar mistake. It can be generated by rule 9 from *them*, but it seems unlikely that *they* would not by itself appear in the top 1'000. The other five extra words might be explained by the fact that Randall Munroe used a kind of spell-checker while writing to help him use only listed words ("As I wrote, I had tools that would warn me if I used a word that was not on the list, like a spell-checker." [10]). This spell-checker seems to have over-generated the said words, perhaps because some of them are morphologically close to allowed words: *mad* to *made*; *hat* to *hate*; and *rid* to *ride*. At other points, Randall Munroe stretches the rules to

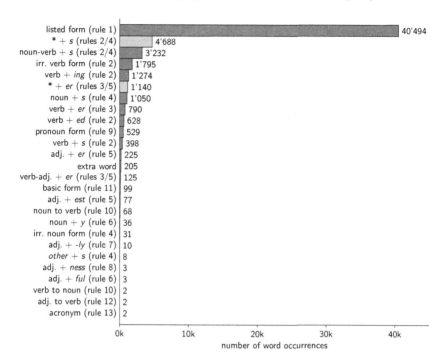

Fig. 2. The origin of the word occurrences in *Thing Explainer* with respect to the different production rules. The total number of word tokens is 51'086.

the extreme. For example, the page about the US Constitution is entitled "The US's laws of the land", even though *US* is not on the list, but the pronoun *us* can be inferred from *we* via rule 9.

Figures 1 and 2 show the distribution of the word forms and word occurrences, respectively, with respect to the production rules. Only about 45 % of the observed word forms are identical to one on the list (i.e. rule 1). If individual word occurrences are considered, about 79 % of the 51'086 word tokens in *Thing Explainer* are directly found on the list. This difference is not surprising, considering that the most common words mostly have just one word form (the top 10 most common words in the book are *the, to, of, a, it, and, in, that, this,* and *you*). Without parsing the grammatical structure of the sentences, it is not always possible to decide which rule was applied, such as in the case of third singular verb forms being indistinguishable from the plural noun forms for words that could be nouns or verbs. Specifically, 260 word forms ending in -*s* have a root word that could be a noun or a verb (e.g. *names*), and 21 word forms ending in -*er* have a root word that could be a verb or an adjective (e.g. *cooler*). Apart from the listed word forms (rule 1), rules 2 to 4 are most frequently used.

As a further note, Arabic numerals were not considered for this analysis. It is not clear whether Randall Munroe wanted them to be part of the language or not. He used them only to refer to page numbers, except for two occurrences of *300* in the phrase "300 years ago", which he could have written out as "three hundred," as he did in many other places.

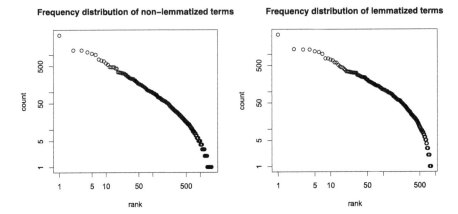

Fig. 3. The frequency distributions of word forms as they are found in *Thing Explainer* (left), and the distribution of "lemmatized" terms when mapped to the list of 1'000 words (right).

2.5 Word Distribution

Finally, we can have a look at the word frequency distribution of word forms as they are observed in the book and "lemmatized" words as they appear on the list. We can check whether they follow a power law distribution (Zipf's law) as closely as other types of texts [6]. There is no obvious reason a priori why a text in a CNL has to follow the same distribution as an unrestricted one, but it would not be surprising either.

The left part of Fig. 3 shows that the distribution of word occurrences follows indeed quite closely a distribution of the kind of a power law, which can be seen by the nearly straight line on the log-log plot. Still the line is curved more than other such word distributions [6], and therefore tends a bit towards a "normal" exponential distribution and away from a pure power law. This effect is even more pronounced in the lemmatized case, as shown on the right hand side part of Fig. 3. For texts in unrestricted language, it has been shown that the lemmatized distribution is normally very similar to the one of plain word forms [7]. Still, both distributions are significantly better explained by a power law distribution than an exponential one (with p-values of 0.022 and 0.017, respectively). We can hypothesize that such kinds of CNL texts in general mitigate the power law effect as compared to texts in full language, but we cannot make any conclusive statement here.

3 Methods

For the analyses presented above, the text is extracted from an electronic version of the book (excluding introduction pages and word list). The book also contains hand written parts, which are not covered. Then all characters except letters, hyphens, and apostrophes are dropped; letters are transformed to lower case; some text extraction errors in the form of missing and extra blank spaces are fixed; words are de-hyphenated; contracted words like *don't* are expanded; *an* is normalized to *a*; Saxon genitive markers *'s* are dropped; the text is tokenized at white spaces; and compound words are split (if the compounds are recognized words). The resulting list of tokens then serves as the input for the analyses. Furthermore, WordNet is used to detect the categories of words and to lemmatize irregular forms. For the power law analysis, the Python package `powerlaw` [2] is used.

4   Discussion

While the topics covered by *Thing Explainer* are entirely serious and the book attempts and (I think) succeeds in seriously conveying complex topics in a highly understandable fashion, the book is also fun and the result of a challenge of the sort *how far can we go*. Randall Munroe admits in the book that "In some places, I didn't use words even when they were allowed. I could have said 'ship,' but I stuck to 'boat' because 'space boat' makes me laugh." At another place, he writes "light drink that wakes you up" and "dark drink that wakes you up," even though both, *tea* and *coffee*, are on the list. On the other hand, there are a number of simple and important words that are not on the list. Randall Munroe reports: "I could have made it easier for myself. There are a few words I was disappointed didn't make the cut. The biggest omission was a synonym for 'rope' or 'string'. [...] the only word I had was 'line'. This fits in some contexts, but has so many other meanings that it was hard to work with" [10]. Other examples where the omission of a word rather leads to confusion than simplification include "the one after eight" for *nine*, "white stuff, like what we put on food to make it better" for *salt*, and "dirt branch" for *root*, apart from the omission of proper names to refer to things like countries or planets. This seems to point to a general problem of such languages with a heavily restricted vocabulary: Writers are forced to circumscribe existing concepts instead of naming them or to involve rough analogies, which can lead to a language that even less precise than full natural language.

These deficiencies could be accounted for to make the language even more useful — at the expense of some of the funniness — by increasing the number of words from 1'000 to, perhaps, 1'500 (like Special English) or even 2'000 or 3'000, or by selecting the words manually (again like Special English) instead of being mainly led by their frequency. There is also a slight inconsistency with respect to how the list of 1'000 words is generated and how it is used. Comparative and superlative forms of adjectives count as separate words when the list is defined, but then these forms can be used in the text even if only the plain form appears on the list. The same applies to adverb forms of adjectives built by *-ly* and pronouns. Normalizing them as well when the list is generated would free some slots for additional words within the limit of 1'000 words.

In general, however, *Thing Explainer* and its language seem to be a huge success, and this success might yield momentum to the general concept of Controlled Natural Language and existing approaches in this field.

References

1. Alderman, N.: Thing explainer: complicated stuff in simple words by Randall Munroe — funny, precise and beautifully designed, 17 December 2015. http://www.theguardian.com/books/2015/dec/17/thing-explainer-complicated-stuff-simple-words-randall-munroe
2. Alstott, J., Bullmore, E., Plenz, D.: powerlaw: a Python package for analysis of heavy-tailed distributions. PloS one **9**(1), e85777 (2014)
3. Alter, A.: Randall Munroe explains it all for us, 23 November 2015. http://www.nytimes.com/2015/11/24/books/randall-munroe-explains-it-all-for-us.html
4. ASD (AeroSpace and Defence Industries Association of Europe). Simplified Technical English, Specification ASD-STE100, Issue 6 (2013)
5. Bourland, D.D.: A linguistic note: writing in E-prime. Gen. Semant. Bull. **32**(33), 111–114 (1965)
6. Clauset, A., Shalizi, C.R., Newman, M.E.: Power-law distributions in empirical data. SIAM Rev. **51**(4), 661–703 (2009)
7. Corral, A., Boleda, G., Ferrer-i-Cancho, R.: Zipf's law for word frequencies: word forms versus lemmas in long texts. PLoS ONE **10**(7), e0129031 (2015)
8. Gates, B.: A basic guide for curious minds, 10 November 2015. https://www.gatesnotes.com/Books/Thing-Explainer
9. Gleick, P.H.: 'Thing Explainer' – a review of Randall Munroe's new book (using the ten hundred most common words), 25 November 2015. http://www.huffingtonpost.com/peter-h-gleick/thing-explainer--a-revie_b_8650772.html
10. Heaven, D.: It's not a rocket it's an up goer. New Sci. **228**(3049), 32–33 (2015)
11. Kuhn, T.: A survey and classification of controlled natural languages. Comput. Linguist. **40**(1), 121–170 (2014)
12. Munroe, R.: Thing Explainer - Complicated Stuff in Simple Words. Houghton Mifflin Harcourt, Boston (2015)
13. Ogden, C.K.: Basic English: A General Introduction with Rules and Grammar. Paul Treber & Co., London (1930)
14. Voice of America. VOA Special English Word Book: a list of words used in Special English programs on radio, television, and the Internet (2009)

Human-Robot Interaction in a Shopping Mall: A CNL Approach

Ezgi Demirel, Kamil Doruk Gur, and Esra Erdem[(✉)]

Sabanci University, Istanbul, Turkey
{ezgidemirel,dgur,esraerdem}@sabanciuniv.edu

Abstract. We introduce a human-robot interaction framework for robots helping/guiding customers in a shopping mall environment. For that, we design and develop controlled natural languages for customers' and robots' questions and instructions. We construct knowledge bases representing general/specific static/dynamic knowledge about shopping malls, to be used in conjunction with the CNLs. We show an application of our framework with a humanoid robot.

Keywords: Controlled natural language · Knowledge representation · Human robot interaction

1 Introduction

In dynamic and complex environments with the presence of humans and robots, human-robot interactions are inevitable. Enabling robots to understand natural language instead of getting the input from graphical user interface makes this interaction more natural. Unfortunately, natural languages are usually more ambiguous and complex than a robot can understand. For that reason, considering specialized domains for human-robot interactions (e.g., in a shopping mall), controlled natural languages (CNLs) can be used to eliminate those ambiguities and allow sentences in a well-defined syntax [12].

With this motivation, we introduce two different CNLs for human-root interactions in a shopping mall environment: H2R-CNL and R2H-CNL. H2R-CNL is used for human-to-robot navigation related questions (e.g., to learn how to get to a store) and imperative sentences (e.g., to guide the human to a store). R2H-CNL is used for robot-to-human multiple choice questions (e.g., between two sorts of stores), requesting some tasks from human when the robot needs help due to its incapabilities (e.g., to push the elevator button), and more importantly answering the navigation related questions (e.g., to describe how to get to a store).

Meanwhile, for the robots to understand the CNL queries/sentences and help/guide humans, we construct special knowledge bases about shopping malls. These knowledge bases include common sense knowledge (e.g., taxonomic knowledge, default knowledge) about shopping malls (e.g., children generally like toys), properties/relations of the shops (e.g., names and locations of stores and what

© Springer International Publishing Switzerland 2016
B. Davis et al. (Eds.): CNL 2016, LNAI 9767, pp. 111–122, 2016.
DOI: 10.1007/978-3-319-41498-0_11

they sell), and temporary properties extracted from observations/interactions (e.g., the east corridor is closed, the main elevator is broken). Some parts of this knowledge, including the taxonomic knowledge and the properties/relations of shops, is represented as an ontology and we call it the shopping mall ontology in the paper. Some part of these knowledge, including default knowledge and the map of the shopping mall, is represented in Answer Set Programming (ASP) [4].

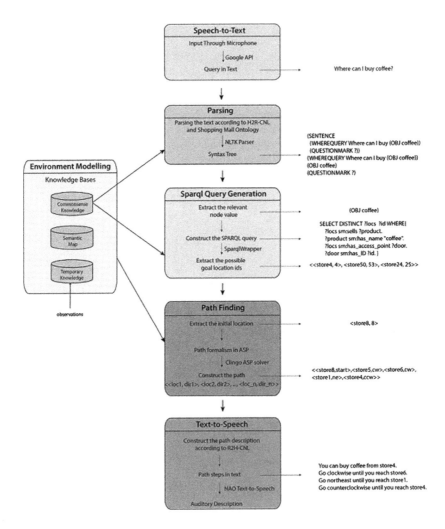

Fig. 1. Program structure

Then, based on these CNLs and the knowledge bases, we introduce an HRI framework (Fig. 1). Our HRI framework has five main components. First, in the Speech-to-Text part, the robot gets the input query or instruction in H2R-CNL

Fig. 2. Human-robot interaction

from the user via speech using Google API [8], and this query/instruction is converted into text format. Then, in the Parsing part, Python's Natural Language Toolkit [3] is used to parse the text according to the H2R-CNL for finding the desired locations or products. After that, a Sparql query [17] is constructed to extract the ids of possible goal locations (e.g., all coffee houses) from the shopping mall ontology. We model the path finding problem in ASP, and then, using the ASP solver CLINGO [6], we can compute paths from the current position of the robot and the customer to the goal location. After computing a path, it is described in R2H-CNL to the user. We illustrate an application of the whole framework with a humanoid robot (i.e., Nao Aldebaran) as shown in Fig. 2.

In the literature, there are various studies on Human-Robot Interaction using natural language [2,9,15,16]. Some of these researches use Graphical User Interface (GUI) for getting the input from user [9], and some of them use speech [2,15] or combine both approaches for getting the users' requests more accurately [16]. Most of these studies focus on only query answering [16] or imperative sentences [2,15]; whereas [9] combines both types of interactions. Some of these studies [2,9,16] use templates to identify query types, but then uses learning-based approaches to identify the specific location or object. In that sense, compared to these related work, our approach is novel in the following ways. First, CNLs are constructed and used for HRI for both human to robot part of the communication and robot to human part. A variety of technologies (e.g., Semantic Web technologies, ASP) are utilized for answering questions and representing the environment in a structured way; for that, different sorts of knowledge (e.g., ontologies, defaults) are integrated. Furthermore, we illustrate an application of our approach on a real robot.

2   Transforming Speech into Text

The efficient transformation of user speech into text that can be parsed is not the main focus of this paper. Hence, the primary goal is to use an external library for speech recognition, developed by specialists focused on this topic.

For that reason, we examined a variety of tools. First we tested the built-in speech recognition capabilities of Nao itself. However, its process of recognizing words primarily depends on searching possible matches within a given vocabulary; meaning a list containing all possible words that can be used during a conversation must be given to Nao before any recognition can be done. Since this is not practical, an alternative use of technology was required.

Many companies, such as Google, have their own speech recognition libraries, which can be accessed with an internet connection. Next, we tested these libraries to decide whether they provide a good solution for speech recognition. The libraries tested belong to Google [8], IBM [11], and Microsoft [14]; where the same set of questions and queries were repeated multiple times in different tones and voice levels in order to draw a general conclusion. On these multiple tries, all of these libraries performed well in terms of response time. However, when it came to understanding the same phrase with differentiating tones and levels, complications started to arise. IBM's Watson Speech to Text required a great amount of enunciation for phrase to be understood, which cannot be achieved in a real time scenario, therefore it has been eliminated. Google and Microsoft's libraries performed well in terms of enunciation and although Google had higher match rate, the difference was not high enough to prefer Google's library over Microsoft's. The verdict was reached when deployment was considered. Since the main concern is easy implementation, API with easier integration would be preferred. Google API required less effort to integrate speech recognition to the system with the use of PyPI SpeechRecognition [19] library; therefore it was preferred over Microsoft's. In the end, speech recognition was realized with the help of Google API due to its easy implementation and high match ratio.

3   Parsing Input Text Using H2R-CNL

After transforming the speech input into text, we used our H2R-CNL to parse it. Our H2R-CNL, as shown in Table 1, supports two sorts of possible sentences relevant to the shopping mall environment: Where-Queries and Imperative-Sentences. Where-Queries support questions like "Where is the women's restroom on the second floor?" or "Where can I buy children's shoes?". The relevant vocabulary needed for these queries are extracted from the shopping mall ontology using the ontology functions $Adj()$, $Loc()$, $Product()$ and $LocCon()$. The extraction of the relevant vocabulary via Sparql queries will be discussed in the next section. Imperative-Sentences can be used for giving some tasks/instructions to the robot. For example, a customer might need a robot's assistance for carrying her packages and may say "Please follow me to the parking lot.". We utilize function $ImpVerb()$ for distinguishing the relevant tasks, e.g., follow and guide. Attempto Controlled English (ACE) [20] does not allow queries like "Where is the women's restroom?."

There are various available tools for parsing an input text according to a particular grammar. In our framework, we use the Python's Natural Language

Table 1. H2R-CNL: A simplified version.

SENTENCE	→	WHEREQUERY QUESTIONMARK
	|	IMPERATIVESENTENCE PERIOD
WHEREQUERY	→	Where can I buy *Adj*() *Product*() ONLOCATION
	|	Where is the *Adj*() *Loc*() FORAGENT ONLOCATION
IMPERATIVESENTENCE	→	Please *ImpVerb*() me to the *Adj*() *Loc*() FORAGENT ONLOCATION
ONLOCATION	→	on *LocCon*()
	|	ε
FORAGENT	→	for *Agent*()
	|	ε
QUESTIONMARK	→	?
PERIOD	→	

Toolkit [3] for parsing the input according to H2R-CNL. This platform is built for Python programming language and it is used for working with natural language data. Since it has a detailed API documentation and it is easy to implement, we use that library in our project as a parser. This tool provides a syntax tree after parsing the sentence in our CNL. Figure 1 provides an example of a sentence and its syntax tree.

4 Sparql Query Generation

Once the syntax tree of customer's query/request is constructed, it is easier to understand what customer wants. For that, we still need to extract the relevant part of the tree, and find the desired location or product accordingly. Since H2R-CNL allows customers to ask a specific place in a region or a specific type of a product, further knowledge about the shopping mall is required. Fortunately, we have a shopping mall ontology.

4.1 Shopping Mall Ontology

Ontologies are formal frameworks for representing knowledge about entities in an environment. Every class of entities is represented as a concept in the ontology. They can be represented as a graph where each vertex represents a concept and each edge represents a relation between the corresponding concepts. For a shopping mall environment, we consider four main concepts: Customer, Product, ServiceProvider and Location. these concepts belong to a super-concept, called *thing*. Sub-concepts are defined via "is-a" relations, such as "Shirt is-a Product." Each concept may have properties (called data properties) that are inherited by its sub-concepts. For example, a product has a property for its "type" (e.g., children's shoes). A specific entity is viewed an "instance" of a concept (e.g., Store X is a French restaurant). Different concepts can be related to each other

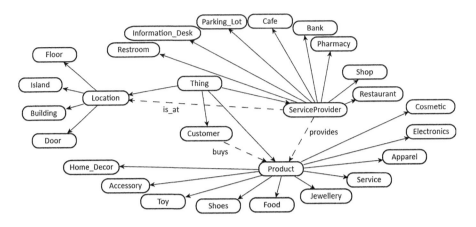

Fig. 3. Shopping Mall Ontology

as well (via object properties). For instance, we define an object property *sell* for explaining the relationship between the service providers and the products.

Figure 3 demonstrates the key concepts and their sub-concepts with solid edges, and the relations between key concepts with the dashed edges. We represent our shopping mall ontology using Web Ontology Language (OWL) [1,10] using the ontology editor PROTÉGÉ [7].

4.2 Constructing Sparql Queries

We use our shopping mall ontology to find the desired locations in the mall, by means of Sparql queries constructed from the syntax trees obtained from parsing of the input query/instruction.

Sparql is a query language for querying ontologies in RDF or OWL. Every Sparql query starts with a SELECT clause for specifying the variables and continues with WHERE clause for specifying the properties of those. Obviously, it would be logical to use an outside tool for constructing these types of queries as well. However the tool we tried, AutoSPARQL [13] required an internet connection with IPv6 capabilities, which would increase the complexity and reduce the feasibility of this component. Hence we decided to construct queries on our own, which was maintainable due to our small sample query space.

Since H2R-CNL allows customers to ask about only locations and products (with some properties), we basically consider two sorts of variables: If the syntax tree contains a LOC node then the customer directly asks about a location; otherwise, she is interested in a product. For example, if the input sentence is "Where is the parking lot for disabilities?," we obtain the following syntax tree:

```
(SENTENCE
  (WHEREQUERY
    Where
    is
    the
    (LOC parking lot)
    (FORAGENT for (AGENT disabilities)))
  (QUESTIONMARK ?))
(WHEREQUERY
  Where
  is
  the
  (LOC parking lot)
  (FORAGENT for (AGENT disabilities)))
(LOC parking lot)
(FORAGENT for (AGENT disabilities))
(AGENT disabilities)
(QUESTIONMARK ?)
```

This syntax tree has a (LOC parking lot) node, so we can start with the location query basis:

```
SELECT DISTINCT ?loc ?id WHERE{
    ?loc sm:has_name"parking lot".
    ?loc sm:has_access_point ?ap.
    ?ap sm:has_ID ?id.}
```

But the customer also specified the type of the parking lot. So we also need to add that information to our query as follows:

```
SELECT DISTINCT ?loc ?id WHERE{
    ?loc sm:has_name "parking lot".
    ?loc sm:has_access_point ?ap.
    ?ap sm:has_ID ?id.
    ?loc sm:for_agent "disabilities".}
```

After the Sparql query is constructed, we can use the ontological reasoner PELLET [18] with this query to obtain the ids of the goal locations.

5 Path Finding

We can represent the "map" of a shopping mall as a graph for the purpose of path finding, In this graph, vertices represent doors and edges represent the accessibility relations between those doors; the labels of those edges represent the qualitative spatial relations (e.g., directions). After modelling the shopping mall as a graph, we can compute a path using also the relevant knowledge bases.

Given a graph G, a knowledge base KB, an initial and a goal vertex, the *basic path finding problem* asks for a path from the initial vertex to the goal vertex in G considering KB. We can obtain the initial position from robot's sensors, and the possible goal positions from the shopping mall ontology using a Sparql query with PELLET. We use ASP to solve basic path finding problem. First, we recursively define the concept of a path by a set of rules in ASP, and represent it in the ASP language ASP-Core-2 [5] so that we can use the ASP solver CLINGO to find a path.

Rules in ASP-Core-2 are of the form `Head :- Body.`, and intuitively read as "`Head` if `Body`." If a rule does not contain `Head` then it is called a constraint, and intuitively read as "`Body` should not hold." If the rule does not contain `Body` then we also drop `:-` from the rule and call it a fact.

We define the vertices and edges of a graph with a set of facts, using atoms of the forms `vertex(U)` and `edge(U,V)`. A path from a vertex `U` to `V` is characterized by atoms of the form `path(U,V,N,X)` expressing that `X` is the `N`'th vertex on a path from `U` to `V`:

```
path(U,V,1,U)   :- vertex(U), vertex(V).
path(U,V,N+1,Y) :- path(U,V,N,X), vertex(X),
                   selected(X,Y), X != V, vertex(Y),
                   vertex(U), vertex(V), index(N), N < maxL.
```

The first rule above expresses that `U` is the first vertex of a path starting from `U`. The second rule expresses that, if there is a path from `U` to `X` of length `N`, then select one of the outgoing edges `(X,Y)` and add it to the path if its length does not exceed the specified maximum length `maxL`. Selection of exactly one outgoing edge of `X` is achieved by the following rule:

```
1{selected(X,Y): edge(X,Y), vertex(Y)}1 :- vertex(X), edge(X,Z), vertex(Z).
```

Then we can define the concept of reachability of a vertex `V` from a vertex `U` in terms of `path(U,V,N,V)`:

```
reachable(U,V) :- path(U,V,N,V), vertex(U), vertex(V), index(N).
```

and add a constraint to express that a goal location should be reachable from an initial location:

```
:- not reachable(U,V), init(U), end(V).
```

To find a path with a small number of locations `N`, we add an "optimization statement".

```
#minimize {N@1,N: path(U,V,N,V), init(U), end(V)}.
```

After computing a path from the initial location to goal location we find the directions using the following ASP rule:

```
path_ex(U,V,1,U,start) :- path(U,V,1,U).
path_ex(U,V,N,X,D)     :- path(U,V,N,X), dir(Y,X,D), path(U,V,N-1,Y).
```

The output of this ASP program is a path with directions:

$$P = \langle \langle V_1, D_1 \rangle, \langle V_2, D_2 \rangle, ..., \langle V_n, D_n \rangle \rangle.$$

6   Describing the Path to User via Speech

We introduce a controlled natural language, called R2H-CNL, for the robot's sentences, as shown in Table 2.

R2H-CNL supports asking Which-Queries when robots encounter an ambiguous situation and wants further details. For example, if the customer asks "Where is the restroom?", than robot needs to ask "Which one do you prefer: for women or for men?". Robots can also request help from humans when they cannot proceed due to its incapabilities. For instance, if a robot does not have a manipulator, then to take the elevator it may say "Could you please push the elevator button to go upstairs?".

Also robots can describe paths, like $P = \langle\langle V_1, D_1\rangle, \langle V_2, D_2\rangle, ..., \langle V_n, D_n\rangle\rangle$, to customers in an understandable and more intuitive way. For that, we decrease the number of steps if it is possible for describing the path more compactly (e.g., if consecutive steps are taken in the same direction) and making it easy to remember. For instance, if the computed path is

$$P = \langle\langle store8, start\rangle, \langle store5, cw\rangle, \langle store6, cw\rangle, \langle store1, ne\rangle, \langle store4, ccw\rangle\rangle,$$

we eliminate the step $\langle store5, cw\rangle$ and describe the path as "Go clockwise until you reach store6. Go northeast until you reach store1. Go counterclockwise until you reach store4.".

Table 2. R2H-CNL: A simplified version.

SENTENCE	→	QUESTION | REQUEST | DESCRIPTION
QUESTION	→	WHICHQUERY QUESTIONMARK
REQUEST	→	COULDQUESTION QUESTIONMARK
DESCRIPTION	→	(Go $Dir()$ until you reach $Loc()$ PERIOD)*
WHICHQUERY	→	Which one do you prefer COLON AMBDISJUNCTION
COULDQUESTION	→	Could you please $Verb()$ the $Obj()$
AMBDISJUNCTION	→	(ONLOCATION or)$^+$ ONLOCATION
	|	(FORAGENT or)$^+$ FORAGENT
ONLOCATION	→	on $LocCon()$
	|	ϵ
FORAGENT	→	for $Agent()$
	|	ϵ
QUESTIONMARK	→	?
COLON	→	:
PERIOD	→	

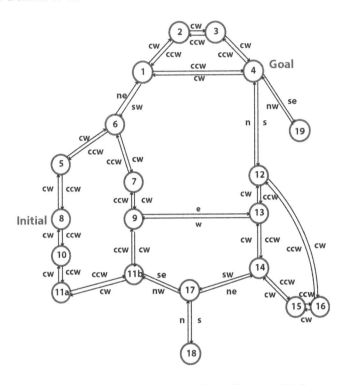

Fig. 4. Graph Representation of the Shopping Mall

7 Case Study

We have implemented our HRI-framework in Python and showed its application with a NAO humanoid robot. Figure 2 demonstrates an interaction example, and Fig. 4 shows some part of the shopping mall graph which includes the described path.

8 Conclusion

We have introduced a Human-Robot Interaction framework for robots and customers in a shopping mall. For that, we have introduced two types of controlled natural languages for more natural interactions: H2R-CNL supports customers' questions and instructions for robot tasks, and R2H-CNL allows mathematical constructs (like paths) to be expressed more naturally. A combination of these CNLs provide a CNL for human/robot-robot dialogue.

Furthermore, for robots to provide requested information and to help/guide customers, we have constructed (i) a shopping mall ontology (and some other knowledge bases) so that they can identify the possible goal locations as requested by the customers, and (ii) an ASP model for path finding in a shopping mall so that they can compute relevant itineraries. We have shown an application of this framework with a NAO humanoid robot.

Acknowledgments. Thanks to Volkan Patoglu for useful suggestions on the design of CNLs. This work is partially supported by TUBITAK Grant 114E491 (Chist-Era COACHES).

References

1. Antoniou, G., Harmelen, F.V.: Web ontology language: OWL. In: Staab, S., Studer, R. (eds.) Handbook on Ontologies in Information Systems, pp. 67–92. Springer, Heidelberg (2003)
2. Bastianelli, E., Castellucci, G., Croce, D., Basili, R., Nardi, D.: Effective and robust natural language understanding for human-robot interaction. In: ECAI, pp. 57–62 (2014)
3. Bird, S., Klein, E., Loper, E.: Natural Language Processing with Python. O'Reilly, Beijing (2009)
4. Brewka, G., Eiter, T., Truszczynski, M.: Answer set programming at a glance. Commun. ACM **54**(12), 92–103 (2011)
5. Calimeri, F., Faber, W., Gebser, M., Ianni, G., Kaminski, R., Krennwallner, T., Leone, N., Ricca, F., Schaub, T.: Asp-core-2 input language format (2013). https://www.mat.unical.it/aspcomp2013/files/ASP-CORE-2.03c.pdf
6. Gebser, M., Kaufmann, B., Kaminski, R., Ostrowski, M., Schaub, T., Schneider, M.T.: Potassco: the potsdam answer set solving collection. AI Commun. **24**(2), 107–124 (2011)
7. Gennari, J.H., Musen, M.A., Fergerson, R.W., Grosso, W.E., Crubézy, M., Eriksson, H., Noy, N.F., Tu, S.W.: The evolution of Protégé: an environment for knowledge-based systems development. Int. J. Hum.-Comput. Stud. **58**(1), 89–123 (2003)
8. Google: Google web speech API (2016). https://developers.google.com/web/updates/2013/01/Voice-Driven-Web-Apps-Introduction-to-the-Web-Speech-API
9. Guadarrama, S., Riano, L., Golland, D., Gouhring, D., Jia, Y., Klein, D., Abbeel, P., Darrell, T.: Grounding spatial relations for human-robot interaction. In: 2013 IEEE/RSJ International Conference on Intelligent Robots and Systems (IROS), pp. 1640–1647. IEEE (2013)
10. Horrocks, I., Patel-Schneider, P.F., Harmelen, F.V.: From SHIQ and RDF to OWL: the making of a web ontology language. J. Web Semant. **1**, 7–26 (2003)
11. IBM: IBM watson text to speech (2016). http://www.ibm.com/smarterplanet/us/en/ibmwatson/developercloud/text-to-speech.html
12. Kuhn, T.: A survey and classification of controlled natural languages. Comput. Linguist. **40**(1), 121–170 (2014)
13. Lehmann, J., Bühmann, L.: AutoSPARQL: let users query your knowledge base. In: Antoniou, G., Grobelnik, M., Simperl, E., Parsia, B., Plexousakis, D., De Leenheer, P., Pan, J. (eds.) ESWC 2011, Part I. LNCS, vol. 6643, pp. 63–79. Springer, Heidelberg (2011). http://dl.acm.org/citation.cfm?id=2008892.2008899
14. Microsoft: Microsoft project oxford SAPI (2016). https://www.microsoft.com/cognitive-services/en-us/speech-api
15. Misra, D.K., Sung, J., Lee, K., Saxena, A.: Tell me dave: context-sensitive grounding of natural language to mobile manipulation instructions. In: RSS. Citeseer (2014)
16. Prischepa, M., Budkov, V.: Hierarchical dialogue system for guide robot in shopping mall environments. In: Habernal, I., Matoušek, V. (eds.) TSD 2011. LNCS, vol. 6836, pp. 163–170. Springer, Heidelberg (2011)

17. Prud'hommeaux, E., Seaborne, A.: SPARQL Query Language for RDF. W3C Recommendation (2008). http://www.w3.org/TR/rdf-sparql-query/
18. Sirin, E., Parsia, B., Grau, B.C., Kalyanpur, A., Katz, Y.: Pellet: a practical owl-dl reasoner. Web Semant. Sci. Serv. Agents World Wide Web **5**(2), 51–53 (2007)
19. Zhang, A.: PyPI SpeechRecognition (Version 3.4) (2016). https://pypi.python.org/pypi/SpeechRecognition/
20. University of Zurich: Attempto (2013). http://attempto.ifi.uzh.ch/site/description/

Erratum to: Controlled Natural Language

Brian Davis[1]([⊠]), Gordon J. Pace[2], and Adam Wyner[3]

[1] National University of Ireland, Galway, Ireland
brian.davis@insight-centre.org
[2] University of Malta, Msida, Malta
[3] University of Aberdeen, Aberdeen, UK

Erratum to:
B. Davis et al. (Eds.)
Controlled Natural Language
DOI: 10.1007/978-3-319-41498-0

The original version of the preface was revised:
The name of an invited speaker as well as several typos were corrected.

The updated original online version for this book can be found at 10.1007/978-3-319-41498-0

© Springer International Publishing Switzerland 2016
B. Davis et al. (Eds.): CNL 2016, LNAI 9767, p. E1, 2016.
DOI: 10.1007/978-3-319-41498-0_12

Multilingual Database Access in the Query Converter

Aarne Ranta[1,2]

[1] Department of Computer Science and Engineering,
Chalmers University of Technology, Gothenburg, Sweden
`aarne@chalmers.se`
[2] Department of Computer Science and Engineering,
University of Gothenburg, Gothenburg, Sweden

Abstract. This paper is a report on work in progress on a piece of software that provides natural language access to databases. The functionalities include translating between texts and entity-relationship models as well as between natural language queries and SQL. The long-term goal of the work is to develop an accurate, generalizable, and scalable technique for natural language access to structured data.

Keywords: Controlled language · Entity-relationship models · Grammatical framework · Query language

1 The Mission

Natural language access to digital data is an old idea, dating back at least to 1959[1]. However, in spite of all efforts, a general, scalable, and intuitive query system in natural language still doesn't exist. The alternatives available for most uses are SQL-like formal notations, hierarchic menus, and string-based search.[2] But the topic of natural language question answering is active again, with systems like Apple's Siri and Google's Voice Search promising to change our practices of information search. Wolfram Alpha[3] is a slightly less general system, but it gives access to sophisticated structured queries in those domains that it covers. The difficulty lies in combining precision with coverage: how to extend accurate query systems beyond limited domains?

In this paper, we will introduce yet another initiative in natural language based data management. It is a CNL (Controlled Natural Language) approach implemented in GF (Grammatical Framework)[4]. Our goal is to develop a *scalable, generalizable, and adaptable approach to accessing structured data*. This includes

---

[1] See W. Woods, "Procedural semantics for a question-answering machine", AFIPS'68, 1968.
[2] SQL was actually meant to be usable by non-programmers and resembles natural language with its COBOL-like syntax.
[3] https://www.wolframalpha.com/.
[4] http://grammaticalframework.org.

© Springer International Publishing Switzerland 2016
B. Davis et al. (Eds.): CNL 2016, LNAI 9767, pp. 123–126, 2016.
DOI: 10.1007/978-3-319-41498-0

- translating between formal and natural language queries,
- adapting the translation to new domains and their concepts with a minimal effort,
- porting translation systems to new languages via lexical annotations.

The work has emerged as a part of qconv, **Query Converter**,[5] an open-source project developing teaching software for an undergraduate database course. The software follows closely the first half of a standard text book.[6] Thus qconv supports exercises and experiments related to textbook concepts: E-R (Entity-Relationship) modelling (graphics and text), functional dependencies and normalization, SQL parsing and interpretation, relational algebra parsing and interpretation, and the XML data format. The idea of qconv is to link all these components together, so that, in particular, natural language access via domain-adapted concepts works across different representations.

The natural language modules of qconv build on earlier work on queries in GF, in particular, on a CNL called YAQL used for natural language SPARQL queries[7]. Working on SQL and the relational data model presents some new challenges, because of its rich hierarchic structure and its varieties of usage and idioms. What we show here is work in progress, yet promising enough to justify a first report and a demo in a workshop.

2   First Results

CNL is currently used in qconv in two places: E-R modelling and SQL queries. Here is an example E-R model with the corresponding text:

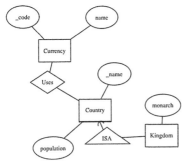

A country has a name and a population.
A currency has a code and a name.
A country can use a currency.
A kingdom is a country that has a monarch.

The CNL is inspired by the mapping between E-R objects and natural language categories proposed by Peter Chen, the inventor of E-R modelling[8] In qconv,

---

[5] https://github.com/GrammaticalFramework/gf-contrib/tree/master/ query-converter/.

[6] H. Garcia-Molina, J. D. Ullman, and J. Widom, *Database Systems: The Complete Book*, 2/E, Pearson, 2008.

[7] M. Damova, D. Dannélls, R. Enache, M. Mateva, A. Ranta, "Multilingual natural language interaction with semantic web knowledge bases and linked open data", in Buitelaar and Cimiano (eds), *Towards the Multilingual Semantic Web*, Springer, 2014.

[8] P. Chen, "English, Chinese and ER diagrams", in *Data & Knowledge Engineering - Special issue: natural language for data bases*, 1997.

Chen's mappings have been slightly generalized so that for instance relationships don't need to be transitive verbs but can be relational expressions more generally. The pedagogical use of the CNL is to serve as an intermediate between an often messy informal text and the formal model. The students are advised to paraphrase the text in the CNL, which makes it easier to draw the model. One could also use the translation in the inverse direction: generate the text from a given model to understand the model better.[9]

The query CNL in `qconv` has a base grammar, which matches the logical structure of queries compositionally. This base grammar is extended with NLG-style (Natural Language Generation) optimizations.[10] Domain-specific terminology for tables and attributes can be inherited from E-R models but also added separately. In particular, it is possible to add domain-specific idioms as a part of a domain lexicon. Here are some example queries and their English and Swedish translations (from a domain that extends the above E-R model).

```
SELECT capital FROM countries WHERE name='Sweden'
what is the capital of Sweden
vad är Sveriges huvudstad

SELECT name FROM Countries WHERE currency='EUR'
which countries have EUR as currency
vilka länder har EUR som valuta

SELECT name FROM Countries WHERE population<1000000
which countries have a population under 1000000
vilka länder har en befolkning under 1000000

SELECT count(*) FROM Countries WHERE population<1000000
how many countries have a population under 1000000
hur många länder har en befolkning under 1000000

SELECT Countries.name, Currencies.name FROM Countries, Currencies
   WHERE continent='Europe' AND currency=Currencies.code
show the country names and currency names for all countries and
   currencies such that the continent is Europe and the currency
   is the currency code
visa landnamnen och valutanamnen för alla länder och valutor
   där kontinenten är Europa och valutan är valutakoden
```

---

[9] This usage is similar to T. Halpin, M. Curland, "Automated Verbalization for ORM 2", LNCS 4278, 2006, which also suggests extensions of the CNL to serve this purpose better.

[10] Technically, the abstract syntax matches relational algebra rather than SQL, and the translation first compiles SQL to algebra in the standard way. One reason is that SQL is a huge language with redundant syntactic structures that have little to do with natural language. Another is that relational algebra permits powerful transformations, normally used in query optimization but often sensible as NLG conversions as well.

As usual in GF, the translation works in all directions, thus both in query interpretation and verbalization.[11] The query translator can also be combined with an ordinary database management system, for instance, via some Java programming using JDBC[12] and GF's Java bindings.

---

[11] The grammar is also used for generating answers, although more work is needed to structure the answers well, especially for speech interaction; cf. V. Demberg, A. Winterboer, and J. D. Moore, "A Strategy for Information Presentation in Spoken Dialogue Systems", Computational Linguistics 37, 2011.

[12] Java Database Connectivity, http://www.oracle.com/technetwork/java/javase/jdbc/.

The Role of CNL and AMR in Scalable Abstractive Summarization for Multilingual Media Monitoring

Normunds Gruzitis[(✉)] and Guntis Barzdins

IMCS and LETA, University of Latvia, Riga, Latvia
normunds.gruzitis@lumii.lv, guntis.barzdins@leta.lv

In the era of Big Data and Deep Learning, a common view is that statistical and machine learning approaches are the only way to cope with the robust and scalable information extraction and summarization. Manning [1] compares Deep Learning with a tsunami at the shores of Computational Linguistics, raising a question if this is the end for the linguistically oriented approaches. Consequently, this question is relevant also to the special interest group on Controlled Natural Language (CNL).

It has been recently proposed that the CNL approach could be scaled up, building on the concept of embedded CNL [2] and, thus, allowing for CNL-based information extraction from e.g. normative or medical texts that are rather controlled by nature but still infringe the boundaries of CNL or the target formalism [3]. It has also been demonstrated that CNL can serve as an efficient and user-friendly interface for Big Data end-point querying [4, 5], or for bootstrapping robust NL interfaces [6], as well as for tailored multilingual natural language generation from the retrieved data [4].

In this position paper, we focus on the issue of multi-document storyline summarization, and generation of story highlights – a task in the Horizon 2020 Big Data project SUMMA[1] (Scalable Understanding of Multilingual MediA). For this use case, the information extraction process, i.e., the semantic parsing of input texts cannot be approached by CNL: large-scale media monitoring is not limited to a particular domain, and the input sources vary from newswire texts to radio and TV transcripts to user-generated content in social networks. Robust machine learning techniques are necessary instead to map the arbitrary input sentences to their meaning representation in terms of PropBank and FrameNet [7], or the emerging Abstract Meaning Representation, AMR [8], which is based on PropBank with named entity recognition and linking via DBpedia [9]. AMR parsing has reached 67 % accuracy (the F_1 score) on open-domain texts, which is a level acceptable for automatic summarization [10].

Although it is arguable if CNL can be exploited to approach the robust wide-coverage semantic parsing for use cases like media monitoring, its potential becomes much more obvious in the opposite direction: generation of story highlights

---

[1] http://summa-project.eu.

B. Davis et al. (Eds.): CNL 2016, LNAI 9767, pp. 127–130, 2016.
DOI: 10.1007/978-3-319-41498-0

from the summarized (pruned) AMR graphs. An example of possible input and expected output is given in Fig. 1.

Article[1]	[..] *An ongoing battle in Aleppo eventually terminated when the rebels took over the city. [..] President Assad gave a speech, denouncing the death of soldiers. [..]*
Article[2]	[..] *Syrian rebels took control of Aleppo. [..]*
Article[3]	[..] *The Syrian opposition forces won the battle over Aleppo city. [..] Syrian president announced that such insurgence will not be tolerated. [..]*

Output Summary:	
Syrian rebels took over Aleppo Article[1] Article[2] Article[3]	*Assad gave a speech about the battle* Article[1] Article[3]

Fig. 1. Abstractive summarization. An example from the SUMMA proposal

While novel methods for AMR-based abstractive[2] summarization begin to appear [11], full text generation from AMR is still recognized as a future task [11], which is an unexplored niche for the CNL and grammar-based approaches.[3] Here we see, for instance, Grammatical Framework, GF [12], as an excellent opportunity for implementing an AMR to text generator.

The summarized AMR graphs would have to be mapped to the abstract syntax trees (AST) in GF (see an example in Fig. 2). As GF abstract syntax can be equipped with multiple concrete syntaxes, reusing the readily available GF resource grammar library, this would allow for multilingual summary generation, even extending the SUMMA proposal.

We assume that the generation of story highlights in the open newswire domain is based on a relatively limited set of possible situations (semantic frames) and a relatively limited set of syntactic constructions (a restricted style of writing) similar to the Multilingual Headlines Generator demo[4] illustrated in Fig. 3.

Although we have not yet implemented a method for the automatic mapping of AMR graphs to AST trees, there is a clear relation between the two representations. From the CNL/GF perspective, the main issue is the open lexicon (named entities and their translation equivalents), however, the AMR *wiki:* links to DBpedia would enable the acquisition of a large-scale multilingual GF lexicon of named entities (as implicitly illustrated in Fig. 2).

With the technique outlined in this paper, the simplified Multilingual Headlines Generator would effectively become the Multilingual Headlines Summarizer with wide applicability in the SUMMA project and beyond.

---

[2] Abstractive summarization contrasts extractive summarization which selects representative sentences from the input documents, optionally compressing several sentences into one.

[3] SemEval-2017 is expected to host a competition on AMR to text generation.

[4] http://www.grammaticalframework.org/demos/multilingual_headlines.html.

Abstract Meaning Representation (AMR)	GF Abstract Syntax Tree (AST)

``
(x3 / take-01
 :ARG0 (x2 / organization
  :wiki "Syrian_opposition"
  :name (n2 / name
   :op1 "Syrian" :op2 "rebels"))
 :ARG1 (x5 / city :wiki "Aleppo"
  :name (n / name :op1 "Aleppo")))
``
	``
	PredVP
	 organization_Syrian_opposition_NP
	 (ComplV2
	  take_over_V2
	  city_Aleppo_NP)
	``

Fig. 2. AMR and AST representations of *"Syrian rebels took over Aleppo"*

Basque	Portuguese	French
Atzerri ministroak akordioa sinatuko du	O ministro dos negócios estrangeiros assinará o acordo	Le ministre des affaires étrangères signera l'accord
Italian	Romanian	English
Il ministro degli affari esteri firmerà l'accordo	Ministrul afacerilor externe va semna acordul	The minister of foreign affairs will sign the agreement
German	Swedish	Latvian
Der Außenminister wird das Abkommen unterzeichnen	Utrikesministern ska skriva på överenskommelsen	Ārlietu ministrs parakstīs vienošanos

Fig. 3. Multilingual Headlines Generator implemented in GF by José P. Moreno

Acknowledgements. This work is supported in part by the H2020 project SUMMA (under grant agreement No. 688139), and by the Latvian state research programmes NexIT and SOPHIS.

References

1. Manning, C.D.: Computational linguistics and deep learning. Comput. Linguist. **41**(4) (2015)
2. Ranta, Aarne: Embedded controlled languages. In: Davis, Brian, Kaljurand, Kaarel, Kuhn, Tobias (eds.) CNL 2014. LNCS, vol. 8625, pp. 1–7. Springer, Heidelberg (2014)
3. Safwat, H., Gruzitis, N., Enache, R., Davis, B.: Extracting semantic knowledge from unstructured text using embedded controlled language. In: Proceedings of the 10th IEEE International Conference on Semantic Computing (2016)
4. Damova, M., Dannélls, D., Enache, R., Mateva, M., Ranta, A.: Multilingual natural language interaction with semantic web knowledge bases and linked open data. In: Buitelaar, P., Cimiano, P. (eds.) Towards the Multilingual Semantic Web. Springer (2014)
5. Ferré, S.: SQUALL: the expressiveness of SPARQL 1.1 made available as a controlled natural language. Data and Knowl. Eng. **94** (2014)
6. Wang, Y., Berant, J., Liang, P.: Building a semantic parser overnight. In: Proceedings of the 53rd Annual Meeting of ACL and the 7th International Joint Conference on NLP (2015)
7. Das, D., Chen, D., Martins, A.F.T., Schneider, N., Smith, N.A.: Frame-semantic parsing. Comput. Linguist. **40**(1), (2014)
8. Banarescu, L., Bonial, C., Cai, S., Georgescu, M., Griffitt, K., Hermjakob, U., Knight, K., Koehn, P., Palmer, M., Schneider, N.: Abstract meaning representation for sembanking. In: Proceedings of the 7th Linguistic Annotation Workshop & Interoperability with Discourse (2013)

9. Lehmann, J., Isele, R., Jakob, M., Jentzsch, A., Kontokostas, D., Mendes, P.N., Hellmann, S., Morsey, M., van Kleef, P., Auer, S., Bizer, Ch.: DBpedia – a large-scale, multilingual knowledge base extracted from wikipedia. Seman. Web J. **6**(2), 2015

10. Barzdins, G., Gosko, D.: RIGA: impact of smatch extensions and character-level neural translation on AMR parsing accuracy. In: Proceedings of the 10th International Workshop on Semantic Evaluation (SemEval) (2016, to appear)

11. Liu, F., Flanigan, J., Thomson, S., Sadeh, N., Smith, N.A.: Toward abstractive summarization using semantic representations. In: Proceedings of the Annual NAACL Conference (2015)

12. Ranta, A.: Grammatical Framework: Programming with Multilingual Grammars. CSLI, Stanford (2011)

Author Index

Printed in the United States
By Bookmasters